1989

University of St. Francis
GEN   781.22 M832
Moravcsik, Michael J.
Musical sound :

W9-DBH-976

Musical
Sound

An Introduction to the Physics of Music

Musical Sound

An Introduction to the Physics of Music

By **MICHAEL J. MORAVCSIK**

Professor of Physics, University of Oregon

Foreword by
ANTAL DORATI

ILLUSTRATIONS BY FRANCESCA MORAVCSIK

A Solomon Press Book

PARAGON HOUSE PUBLISHERS

LIBRARY
College of St. Francis
JOLIET, ILLINOIS

Copyright © 1987 by Michael J. Moravcsik

All Rights Reserved

No part of this book may be reproduced in any form by any method, either in existence or yet to be devised, without the written permission of the publisher. Inquiries should be addressed to:

Paragon House Publishers
866 Second Avenue
New York, NY 10017

The author and publisher wish to thank the copyright owners who have given their permission to use copyrighted material. Any omissions or errors in giving proper credit are unintentional and will be corrected at the first opportunity after the error or omission has been brought to the attention of the author or publisher.

MUSICAL SOUND: An Introduction to the Physics of Music is a joint publication of PARAGON HOUSE PUBLISHERS and THE SOLOMON PRESS. It is distributed by PARAGON HOUSE in the United States of America and by FEFFER AND SIMONS outside the U.S.A.

Library of Congress Cataloging in Publication Data

Moravcsik, Michael J.
   Musical sound.

   "A Solomon Press book."
   Bibliography: p.
   Includes index.
   1. Sound. 2. Music—Acoustics and physics.
I. Title.
QC225.15.M65   1987   781'.22   86-30278
ISBN 0-913729-39-6

Book Designed by Reuven Solomon

Manufactured in the United States of America

781.22
M832

Table of Contents

133,459

List of Figures

LIST OF TABLES

LIST OF PHOTOGRAPHS

Acknowledgments

In registering indebtedness in connection with a book, one faces the question of how far back to go. Science and music have always constituted the two most important facets of my life, and the list of people and circumstances that made this possible, encouraged it, and contributed to it is too long to enumerate. Turning to more specific influences, my direct interest in the science of musical sound arose when I was asked to teach, at the University of Oregon, a one-term course, by then well established mainly due to the efforts of a colleague of mine, Kwangjai Park, called The Physics of Sound and Music, catering to a substantial extent to students in the university's distinguished School of Music.

As my decision matured to write a textbook for such a course and for the many music-lovers "at large," I was greatly encouraged by the ready cooperation of Sidney Solomon, of Publishers Creative Services, with whom I had a happy association in connection with my previous book on a quite different subject.

Three members of my family also played a very substantial role in this book. My sister, Edith Moravcsik, professor of linguistics at the University of Wisconsin in Milwaukee, volunteered to be a guinea pig on whom I could try out the manuscript. Not at all a natural scientist but an avid music lover, she went over the book with a fine tooth comb and made a long list of suggestions which were incorporated in the final version. My wife Francesca, a landscape architect, took time out to prepare the illustrations from the sketches drawn by me, a born antitalent in drawing. Finally, my daughter, Julia was not only instrumental in preparing the manuscript and the index but also made stylistic and other suggestions which improved the book.

Various earlier versions of the manuscript were also used by students in several of my courses on this subject, and their complaints and praises were also a beneficial influence. In a similar vein, I hope that complaints and praises from the readers of this book will be forthcoming in order to continue to improve it in subsequent editions.

MICHAEL J. MORAVCSIK
Eugene, Oregon

Foreword

W HAT IS MUSIC?

Where does it come from?

What does it do?

What does it mean to us?

Why does music belong to human life?

These are the main questions that have provoked—and continue to provoke—the huge literature around that sublime form of art. The more answers are attempted, the more questions are raised.

Arriving, through manifold and complicated channels, to an axiom-like statement, it can be said that music represents, mirrors, sublimates the three basic fields of human emotion and intellect: love (sensuous and spiritual), by melody; work (activity) through rhythm; community, (togetherness, brotherhood) with harmony.*

---

*Responding to possible arguments:

Music can depict, of course, all the above also in the negative: hatred, inertia, and anarchy—but it does so only to point out contrasts. Its goal always is to lead human thinking and feeling into positive channels so as to bring light and freedom.

Mozart—whose intuitive musicianship is unequalled—wrote once to his father: "Music shall never offend." How right he is. Real music always is a source of consolation and hope to those who understand and feel it, no matter how hard it can be to arrive at that platform. Music *gives* abundantly, but also *demands* from both its makers and receivers.

The finality of this postulate could make us think, that with it we have reached the ground.

But we have not.

Not by far.

The next question is right at hand: How does this representation take place? By what means is music capable of penetrating such depths of human existence?

Through sound, of course.

We all know that.

But what is sound?

Yes—what is it? That infinite variety of voices that reach our consciousness through our ears, that our brain digests or rejects, that surrounds us constantly, day and night—including silence, which is a sound also—and what is that sparsely selected group that we call musical sounds—a small number of sounds, a number slowly mounting through the centuries, now about two hundred—two hundred!—taken out at random—random? What random?—out of the billions that float around, to make our art of music with them—what is all that? How does it come about?

With this question we have arrived at the book of Prof. Moravcsik, which these lines are to introduce.

As a musician who is active in the creative as well as in the performing realms of music, I have read this book, slowly and carefully, often re-reading a paragraph or a sentence, not only with interest, but with mounting fascination.

Its contents were only fractionally included in my early education. Much was entirely new to me, although I know of the existence of all that information. The *terrain* was familiar, but only that.

I had the feeling of entering the cellar of the house in which I live and know intimately, from the ground floor upward. Not that I wasn't in that cellar before. But when I went down there, it was only for a few moments and with a definite purpose, to bring up a suitcase, a bottle of wine, or whatever I needed among the multitude of objects stored in that subterranean territory, dark and slightly mysterious and distant. I have never "wandered around" in that cellar before.

Now that I have, I can happily report that I came up again, enriched by an important experience—and intact. The excursion did not change my musicianship, the information gathered did not alter my musical attitudes, methods, goals, ideals. But I became richer and pleased, contented by having been told these "facts of life" of the art that I am serving—and told in such clear, in such— the word that is on the tip of my pen seems unavoidable— *natural* terms.

The "facts-of-life" analogy that suggested itself, is—I believe— rather apt. Indeed, the knowledge of them, in the realm of sex, where the term is conventionally used, does not have a negative ef- fect upon the spiritual "ingredients" of sex-life. Love, passion, ex- altation do not suffer from our knowing the "facts-of-life," and the poetry of love does not lose its impact upon those who know these facts.

This book—as its author emphatically states right on the first page—is not intended for scientists. That promise is rigorously fulfilled until the last page is turned.

It is good so.

These kinds of books should exist—and I hope they do—on every subject.

Everybody is ignorant—about something. And it is very impor- tant that we should be getting information in the various fields of our ignorance. It is also very important that those who undertake to disseminate needed information take into account that adult laymanship that they undertake to address and not be "blinded" by their own knowledge.

Thus, in the case at hand, it is refreshing that laymen are told in laymen's terms (but attention! what we hold in our hands is by no means, a juvenile transcription or a reader's-digest edition) about the sources of music, and only as far as the sound itself is concerned: its origin, its mechanics, its "bodily" existence, and told by an expert scientist who has set his own limits and does not wish to trangress them.

One important qualification must be interpolated here: when the author says that his book is for lay readers he means laymen in physics. The reader's relationship to music is not his concern. He obviously would regard musical interest as an asset, but his book is wide open for readers who are not musical but are interested in acoustics for other reasons. (The enormous advances in that field, that are being made in our days, open—among other things—vast areas of laymanship and dilletantism as well.) It is very possible that someone who picks up this book out of general curiosity only, will end up by reading it as a hi-fi-fan or a music-enthusiast. The author's own position is clearly defined, although he does not waste any words about it: it is that of the musically interested scientist. Perhaps there was in his early youth a moment in which he vacillated between a musical or scientific profession, but ever since he made his decision he stands by it and never goes beyond the goals he has set for himself. In his book he does not treat the art of music, but the means, the basic "tools" of that art.

The line is neatly drawn—and this is for the good. There are no "sour grapes" in this book, because the hand is stretched out only to those it can reach.

An example to the contrary comes to mind. A very eminent colleague of mine, a great and widely esteemed musician—my elder of nearly two decades—wrote a thoughtful, interesting, searching book in which he endeavored to explain the evolution of the art of music on the basis of its physics only. (He came to a pessimistic conclusion.) During a debate, discussing that book with its author, the writer of these lines, then a young man (whose attitude concerning the same problems was—and still is— optimistic) was prompted to exclaim: "But, my dear friend, you have described and discussed every facet of music in your book, omitting only the single one that you were longing to write about— the talent for it!"

Professor Moravcsik did not fall into that trap. His book is not concerned with musical talent, whether creative, recreative, or receptive (let us not underestimate the overall importance of the last mentioned!)—even with music itself he is only concerned as far as the sound it makes.

He well knows, however, that the impact of music is beyond its sound: it is in the region of associations, fantasies, emotions kindled by those sounds in that most mysterious human "organ"— the word is used in want of a better one—the soul—that has no physics.

This region Moravcsik does not enter. But he leads us to its door, and we shall be thankful for that.

—ANTAL DORATI

Musical Sound

Sound

An Introduction to the Physics of Music

Why and How—An Explanation for the Reader

T HERE ARE AT least a dozen books on the market on the physics of sound and music. Most of them tend to be encyclopedic and mathematical: they provide an assortment of often unexplained factual information, and the exposition is interlaced with formulas. They are useful as reference books for certain facts, and also to the scientifically trained who know some physics and can handle algebra.

This book is different, for at least two reasons. First, it concentrates on getting across ideas and concepts, thus building a base that allows the easy perusal of specialized reference books. Second, it is aimed at those with no background in science, and with only an elementary knowledge of mathematics, that is, some fluency in the four basic arithmetic operations applied to numbers. No algebra is used, no trigonometry is required. What little math is used is reviewed in an appendix for those who have become a bit rusty since their days in grammar school.

As a result, the book is aimed at two broad groups of people.

First, it is a textbook for a one-term or one-semester course for students who are not concentrating on science. If such students are afraid of science and claim to have zero ability to handle math, they need not worry when reading this book. Science here will be quite different from what one might expect in the usual science course and no math will impede progress toward understanding the concepts. Indeed, this book grew out of my having taught a "Physics of Sound and Music" course many times and being dissatisfied with the various textbooks I used. And my students were dissatisfied, as well.

In particular, we were frustrated with the available textbooks for three reasons. First, they were, on the whole, much too lengthy for a one-term course, and hence could not really be used as a textbook but only as a reference book. Second, their mathematical prerequisites were considerably more severe than the background of the students participating in the class, and this mathematical underbrush served as an obstacle for the students to concentrate on the physical understanding. Third, many of the statements in the available books were of an encyclopedic type, listed without explanation. Although appropriate for a reference work, such unexplained elements are confusing in a text book.

The second group of readers to whom the book will be of particular interest is to be found outside the classroom setting. They are the music lovers, phonograph record buffs, amateur and professional musicians, hi-fi enthusiasts, and other sound-related activists. Many of them will have had an urge to understand something about sound and music from a scientific point of view, but, as the saying goes, were always afraid to ask, because the available texts were much too technical and ponderous for easy reading. This book, on the other hand, requires no previous knowledge of the subject or no preliminary work. Rather, basic ideas and concepts are emphasized in an easy-to-read context.

Here are some suggestions as to how to use the material to best advantage.

The scientific investigation of a problem consists of several steps. First, we have to understand the concepts, the principles, the ideas of structure that govern the phenomena involved. The second step is to be able to acquire an approximate feeling for the quantitative aspects of the problem. For this, only the basic arithmetic operations that one uses every day are needed. The third step is to carry out a detailed and quantitatively rigorous mathematical analysis of the problem. For this a considerable amount of physics and significant mathematical sophistication are needed, and hence *in this book we do not do that step.* It is the unnecessary mixing and confusion of this sophisticatedly analytic part of science with the concepts and approximate quantitative descriptions that have made some science textbooks for nonscientists unpalatable for the student.

But there is no such thing as a free lunch, and to benefit from this book, the reader must spend some time and effort. I recommend that this effort be in two directions. First, you should use your intelligence, imagination, and reasoning power to understand and internalize the ideas, concepts, and explanations we discuss. Then, in order to test yourself, you should work out as many of the problems at the end of the book as you can. It is a unanimous verdict of scientists and science teachers that only through problem solving can one assimilate the knowledge that one acquires. The problems given in this book are very simple mathematically (utilizing only the four elementary arithmetic operation), so that if you cannot do the problems, do not blame it on the math. The source of the trouble is in your not understanding the physical ideas. Solutions to the problems are also provided, but you should not look at these until you have made a serious effort to work out the problems yourself.

Once having put a fair amount of work into perusing this book, what will you gain from it? On a utilitarian level, you will acquire a sound conceptual and semiquantitative feeling for the fundamental scientific principles behind music—the way it is constructed, produced, performed, listened to, transmitted, and stored. This will be helpful professionally if you are active in anything related to music, and will allow you to consult the specialized books more easily. Indeed, this book may encourage you to plunge deeper into the study of acoustics and other scientific aspects of music. A list of references is given to accommodate you in this.

On an esthetic level, you may enjoy the newly acquired understanding, and it may give you a new insight into how science works, on a less formal level. It is important that science shed the awesome and esoteric aura that now surrounds it, and that it mean something to each person who comes into contact with it. Science is only one way of looking at the world, but it is a powerful one that can be applied in everyone's daily life.

So relax, and enjoy the reading. Be confident that you can master this book: it was written for you.

A few words are in order about the book's structure. It consists of five main parts. The first introduces you to some of the physical concepts that play a crucial part in the discussion of music. This is followed, in the second part, by a description of how sound travels from the source to you and how you perceive the sound. In the third part, we discuss the characteristics of the type of sound called music. The fourth part deals with how we produce musical sound through instruments and the human voice. Finally, in the last part, we turn to the enviroment in which we listen to music, and to the tasks of transmitting music over space and time. The progression is a logical one, and so the reader is advised to start at the beginning and read through to the end, although chapters are understandable on their own, without the benefit of the previous chapters.

Part I
The Concepts of Motion, Vibration, and Waves

Prologue

"IT IS EASY to explain what happens when Segovia gives a recital in a concert hall to a live audience and the recital is also broadcast over the radio and taped at the same time," said the scientist. "He exerts force on the string and displaces it from its equilibrium at-rest position, thus giving the string potential energy. When his finger releases the string, a restoring force acts on it, and the string acelerates and picks up speed, converting its potential energy into kinetic energy. When the string passes through the rest position, its potential energy is zero, its kinetic energy maximum, its speed also maximum, and its acceleration zero. Then, as the string overshoots, the kinetic energy is converted back into potential energy, the acceleration is negative, and the speed lessens, until the string lies about as much on the other side of the rest position as it did on the first side. But this is not quite true, because the second law of thermodynamics forces some of the kinetic and potential energy to convert into heat and so the vibrations of the string are dampened. Meanwhile the vibrations are also transmitted into the body of the guitar, where resonances respond to the vibrations, and the air in the guitar, as well as the plates of the instrument, will also vibrate with the same frequencies. The vibrations are transmitted into the air, and create longitudinal sound waves, which then propagate with an intensity obeying the inverse square law, except that there are reflections from the walls of the auditorium, governed by the absorption coefficient of the walls, and the many sound waves form interference patterns. Then the sound reaches the ear of the listener, where it is converted, by the well-known mechanism of the ear, into nerve pulses going into the brain. But the sound from Segovia's guitar is also fed into the microphone, in which it is converted into electric current and then, by the antenna, into electromagnetic waves, which reach radio sets far away. Also, the current of the microphone is used to run an electromagnet that creates magnetic patterns on the tape . . . "

"Stop," I said, "You are far ahead of me. You have been using expressions that have only a vague meaning for me, if any at all. You have to give me a little time to read the first six or so chapters of this book so that I catch up on a bit of elementary physics and thus understand your terminology and concepts. Then I will let you continue in this vein—and you *will* communicate to me. Now you do not."

Chapter One
About Motion

IN DISCUSSING SOUND, music, hearing, or the structure of musical instruments, we will be focusing mostly on motion and waves (which also involve motion), and so it might be appropriate to review what we need to know about motion in general. The well-developed branch of physics called mechanics deals successfully with the qualitative and quantitative description of all sorts of motion in a very sophisticated and highly mathematical way. We will not need all that powerful and beautiful formalism in our attempt to understand the basic physical ideas of sound and music, but can rely exclusively on a few aspects of motion that are familiar from everyday life. So just relax and recall you own personal experience with motion—an experience that is enlarged almost every minute of your life.

Our discussion of motion will be in two parts. First, we develop concepts that help us to *describe* motion. These are location, velocity (and speed), and acceleration. In the second part, we introduce concepts that help to *explain* motion. These are force and energy. In connection with the latter, we will acquaint ourselves with the most universal law of nature, that of the conservation of energy, and with another universally applicable law that describes the way energy is converted from one form into another.

Describing Motion

Location

At any given time, we can (partially) characterize something by giving its location, that is, its position with respect to some agreed-on point of reference. You give instructions to the lost motorist by indicating that three blocks ahead he will have to turn right and drive 1 ½ more blocks to find, on the left, the movie house he is seeking. In this case the location can be described in two "dimensions," that is, by specifying how many more blocks he has to drive straight ahead and then how many blocks he has to drive in a perpendicular direction. In this example blocks constitute the unit of measurement, but you also could have told him to drive 1200 feet straight ahead and then 640 feet to the right, thus giving the location in terms of different units.

In a more general example, the location will have to be specified in terms of three numbers (ahead or behind, right or left, and up or down), since we live in a three-dimensional world. (In the foregoing example, up or down does not enter into our consideration since we move on a plane surface, the surface of the earth.) These three spatial coordinates can then specify the location of any object. In other words, the location has a direction also, not only a magnitude. We will see that the same is true for our other concepts used to describe motion.

We call motion a situation in which the spatial location of an object is, in at least one of the three dimensions, different at different times. Indeed, motion can be described completely in terms of a table that specifies the three spatial coordinates of the object under study at all different times.

In the following we will discuss some other concepts and quantities describing motion, which are very useful in getting a quick idea of the characteristics of the particular motion with which we deal. These new concepts are speed, velocity, and acceleration. It should be noted, however, that these additional concepts are all secondary, derived from location and its variation with time, and thus they contain no more information than location itself at various times. The only thing they do is provide the information in a form that is handier for certain purposes. This point will be evident from the example in Chapter 2, where, by giving the location of a bus at various times, we will be able to determine its speed and acceleration also.

Speed and Velocity

In particular, in everyday life we use the word "speed," which tells us how much change in location occurs in a given time. To compare various speeds easily, we usually refer to a unit of time (hour, minute, second, etc.), and give the amount of change in location that occurs in that unit of time. The change in location is given by the distance between the location at the beginning of the time period and that at the end of the time period. Distance is measured in its own units, which may be miles, feet, meters, centimeters, millimeters, and so on. Thus speed comes out to be in units of distance per time, for example, in miles per hour or meters per second.

By giving just the speed, we provide no information on the *direction* in which the motion proceeds. Such additional information is, however, often useful, and so it is also convenient to use another concept call *velocity*, which is speed together with the specification of the direction in which the motion takes place. For example, a motorist who starts from Eugene, Oregon, and drives 50 mi/h would be hard to locate since we have no idea in which direction he is heading. Only his speed is given. If, however, we say that his velocity is 50 mi/h in the northerly direction, it seems we have specified his motion completely.

Or have we? What if he really moves in a northwesterly direction (Figure 1.1), and *as far as his north-south position is concerned*, is 50 miles farther north every hour? That situation also fits the description that his speed is 50 mi/h in the northerly direction, but the description should not end there, since we should also tell in what way his location changes in the east-west direction. If we also specify that his velocity is 40 mi/h in a westerly direction, we have indeed specified his motion completely. We have given his speed *along two different dimensions*, and thus (in the case of a motorist without wings) we have described his velocity completely. (If the motion is in three dimensions, we need to give the speed in three directions.) We call this process the specification of the *velocity components*: the north-south components (50 mi/h in the northerly direction) and the east-west component (40

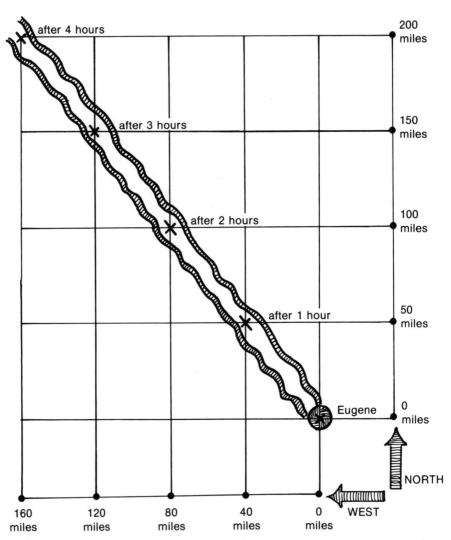

Figure 1.1   The location of a car that travels from Eugene, Ore., at a speed of 50 mi/h in the northerly direction and 40 mi/h in the westerly direction.

mi/h in the westerly direction). Incidentally, if the motorist actually proceeds along a straight road in that northwesterly direction, his speedometer would show neither 50 mi/h nor 40 mi/h, since he would have to go faster than either of these speeds to make simultaneous progress by 50 mi/h northward *and* 40 mi/h westward. His actual speed as shown on the speedometer would be about 64 mi/h, but how we calculate this need not concern us here, as long as we understand that the speed in this composite direction must be larger than either of the two components.

Acceleration

To explain the last concept that we need in connection with our everyday description of motion, let us consider the following situation. Only seven minutes are left before the movie starts, and so you jump into your car and step on the accelerator. The car takes off and speeds up rapidly. While a few seconds ago it was not moving (its speed was zero), now it travels at 45 mi/h. During these few seconds, therefore, its speed did not remain constant but increased all the time, as the needle on the speedometer demonstrated. The *rate* of the change of the speed is described by the *acceleration,* which says how much the *speed* changed in a unit of *time.* Thus the acceleration has the same relation to speed as the speed had to location: In each case the first tells how much the second changes per unit time. It is then evident that the unit of acceleration is speed per time, just as the unit of speed is distance per time. We can, for example, measure acceleration in "(miles per hour) per hour," denoted also as miles per hour squared, as if you could square an hour. You cannot—this is just a notation.

The word acceleration comes from the Latin *ad*-celeratio, where "ad" means "adding to." Thus we use acceleration when the speed *in*creases. When the opposite happens and the speed *de*creases (e.g., when you step on the brake of the car), the resulting pattern is *de*celeration (from the Latin "de" meaning, in this case, subtracting from).

Just as in the case of speed, the complete specification of acceleration or deceleration also requires the specification of the direction. This involves giving all three components of the acceleration or deceleration—up or down, right or left, and forward or backward.

It is regrettable that in the case of speed and velocity we had two different words to describe the magnitude alone and the magnitude and direction together, whereas in the case of acceleration we have only one word to describe both these concepts.

As mentioned earlier, speed, velocity, acceleration, and deceleration are *secondary* quantities in that they can be deduced from a table giving the location at all different times. Some of the exercise problems for this chapter demonstrate this.

Understanding Motion: Force and Energy

The two concepts we now discuss are not primarily used for *describing* motion, but for *understanding the causes* of motion. They are also words we commonly use in everyday life. We may make the meanings of these two words a bit more precise so that they can serve science better, but the basic ideas remain.

Force

Force is very commonly used in everyday parlance. If you lean against and exert force on a filing cabinet in the middle of your office, it will move. The more force you apply, the faster it starts moving (i.e., accelerates). Force is also needed for deceleration. When your car crashes into a wall, the wall exerts force on the car, quickly decelerating it. In the rigorous usage of physics, force is visualized as the cause of any motion with acceleration or deceleration. Indeed, we say that the acceleration or deceleration is proportional to the force, which means that for a given object, *twice* as much force applied to it will produce *twice* as large an acceleration, and so on.

Note that force is thus related to acceleration and not to velocity or speed. If you exert force on a bowling ball and impart acceleration to it (by taking a ball at standstill and getting it rolling), from then on the ball will roll with uniform speed without your touching it, and thus without your exerting more force on it. (To be sure, that uniform speed will be larger if you used a larger force, but that is so because you caused a greater acceleration, resulting in this larger speed.) Thus to *continue* to have the *same* speed requires no force.

Although force is proportional to acceleration or deceleration for a given object, this relationship will be different for different kinds of objects. If you apply a certain amount of force to a filing cabinet, you will produce more acceleration or deceleration than if you apply the same amount of force to a large car. Clearly the acceleration or deceleration produced by a given force also depends on the "amount of stuff" you have to move. This "amount of stuff" is called "mass" in scientific language. On earth, the mass can be made equivalent to weight so we can say that a given amount of force produces less acceleration or deceleration on a heavy object than on a light one. Indeed, force is proportional both to mass and to acceleration or deceleration. This will be obvious if you think about it in terms of everyday experience. Parenthetically, however, mass can also be measured independently of the concept of weight, which is good since it assures that most of what we will learn about the scientific basis of music also holds in spaceships in a gravitation-free enviroment where everything is weightless, but not massless.

Energy

We now turn to the other concept we will need, namely, *energy*. This also comes from everyday life. An energetic person is somebody who works hard over a long period of time. At least one of the scientific meanings of energy comes from that concept. We say that an energetic person moves quickly. We will also incorporate this into the scientific concept of energy.

But the idea of energy is broader in science than in everyday life. We have several kinds of energy in science. One is associated with motion, and is therefore called *kinetic* energy. This comes from the Greek word "kinetikos," meaning associated with motion; the word also sired "cinema" for motion picture. We say that a given object has kinetic energy if it moves, that is, if it has a speed. In fact, for reasons we need not be concerned with here, kinetic energy is defined as being proportional to the *speed squared*. This means that a given object moving four times faster would have four times four, or 16 times, more kinetic energy. You can see that this proportionality of kinetic energy to the square of speed is the link between energy and motion. Whenever there is motion, we can define the kinetic energy that is involved in that motion. Through the law of conservation of energy (which we will discuss presently) some important features of the motion can be specified if we can determine the kinetic energy in that motion. We will see an example of this when we discuss the vibrating string later in this chapter. We also say that energy is proportional to the mass of the object, so that an object that is twice as heavy would have a kinetic energy that is twice as much.

It is useful, however, to define other kinds of energy that are not so directly related to the everyday meaning of the word. Consider, for example, a huge boulder suspended above a table. Nothing moves, and yet we have an ominous feeling that except for the ropes holding the boulder (something that we can eliminate with practically no effort), there would be plenty of motion. The boulder would come crashing down, smashing the table, damaging the floor, and arriving at the lowest point of its path with a considerable amount of speed, and hence kinetic energy. We can describe the suspended boulder as having a considerable amount of *potential* for acquiring kinetic energy. We thus say that the boulder has *potential energy*, which is converted into kinetic energy as it falls and keeps losing its potential energy.

Apart from the intuitive sense of this definition, it also turns out to be useful because, by working with the two concepts of energy together, we can formulate a general principle called "conservation of energy." Here "conservation" means something distinctly different from the energy-conservation measures related to insulating a home. The scientific meaning of conservation is that the amount remains the same for a given system as time goes on. Thus conservation of energy means that the total energy (kinetic energy and potential energy added together) remains the same as time goes on. Such a rule turns out to be very generally true for any isolated system in the world, with one additional remark to be made in the context of sound and music. This is that energy connected with motion can be either that of an everyday object as a whole (what we usually call kinetic energy), or that of the individual molecules inside this object in a disorganized, uncorrelated way, in which case we call it "heat." Thus if we look at the conservation of energy from a practical, standpoint, we can say that the total energy (consisting of kinetic energy, potential energy, and heat energy) must be the same for an isolated system (that is, if energy is not lost to the outside or not pumped into the system from the outside), regardless of when we view this system.

In contrast, the everyday meaning of energy conservation is that you want to use altogether less energy by applying more of the energy toward your purposes and letting less go to waste. Whether you "conserve" energy in the everyday sense or not, energy conservation (in the scientific sense) holds.

Energy Conversion

Within the overall amount of energy, however, we can have transformations of energy from one form to another. Indeed, we can often get a very deep and revealing insight into the working of things in nature by just following the energy-conversion processes that take place. In our context, and anticipating a bit, consider (Figure 1.2) a tight string being plucked with your hand. As you exert force on the string and pull it out of its equilibrium position, you work hard to impart some potential energy to the string. As you hold it away from its straight, unplucked position (Figure 1.2A-a), the string has potential energy, but no kinetic energy, since it is at standstill. It may have some heat energy in it (due to the motion of molecules, which always occurs at room temperatures), but not significantly more than it had in its original, unplucked state. You then release the string. The force of its tension pulls the string toward its original equilibrium position (Figure 1.2A-b), and it quickly loses its potential energy (by approaching the equilibrium position), but at the same time it picks up speed since the force of tension keeps acting on it, and hence is increasing its kinetic energy. When it has reached the original equilibrium position (Figure 1.2A-c) it has zero potential energy, but it is going quite fast, and thus has much kinetic energy. It therefore overshoots, and proceeds toward a "plucked" position at the opposite side. In doing so, the tension of the string works against its motion and produces a deceleration in its speed. Thus the string, as it regains potential energy, keeps losing kinetic energy (Figure 1.2A-d). After awhile, the string stops (Figure 1.2A-e) (now having no kinetic energy but lots of potential energy), and turns around, moving toward the equilibrium position again, and so on. This goes back and forth (Figures 1.2A-f–i), and would do so indefinitely if there were no loss of energy (1) to the air around the string with which the string constantly collides as it moves back and forth, and (2) to the string itself in the form of heat; that is, as the string bends back and forth, it keeps heating up. Thus the *sum* of the kinetic energy of the string and the potential energy keeps decreasing slowly, with the difference going into heat and energy in the air. At the end all the energy has dissipated into heat. If, however, we consider the *total sum* of kinetic energy of the string as a whole plus the potential energy of the string plus the energy transmitted to the air plus the heat generated in the string, this total sum always remains the same no matter when we look at the system.

The description of this process of a vibrating string in terms of location, speed, and acceleration is shown in Figure 1.2B. Energy conservation is a very general law of nature, although in more complicated cases other forms of energy also need to be considered. For example, potential energy can result not only from a particular position of an object (as in the case of the suspended boulder), but also from a particular chemical composition (for example, an as-yet-unexploded stick of dynamite). In a different context, in nuclear reactions mass can be converted into energy, and so the "mass equivalent of energy" must also be included in the energy accounting. For our purposes, however, these more esoteric aspects of energy conservation will be irrelevant, since very few musical instruments operate with dynamite or nuclear power.

One last remark needs to be made in connection with energy-conversion processes in which one form of energy is converted into another. It turns out that some forms of energy are more likely to be *converted into* than others. In particular, heat is the form of energy *into* which processes like to convert preferentially. This preference can be perceived only on the average, but it eventually wins out over other conversion processes and the energy of the system ends up entirely in heat. This is evident in our example of a plucked string, in which the end result is the string at rest again with no potential or kinetic energy, and heat in both the air and the string. What heat has over other forms of energy is its *disordered* character; that is, in heat the energy is distributed widely among the huge number of molecules of the object, with each molecule moving in some direction unrelated to the direction of the next molecule. In contrast, the kinetic energy of the string as a whole is much more ordered, in that all molecules of the string move in the same direction and at the same speed—which is what we mean by "the string as a whole moving." We thus can say that energy-conversion processes in nature tend to go from greater

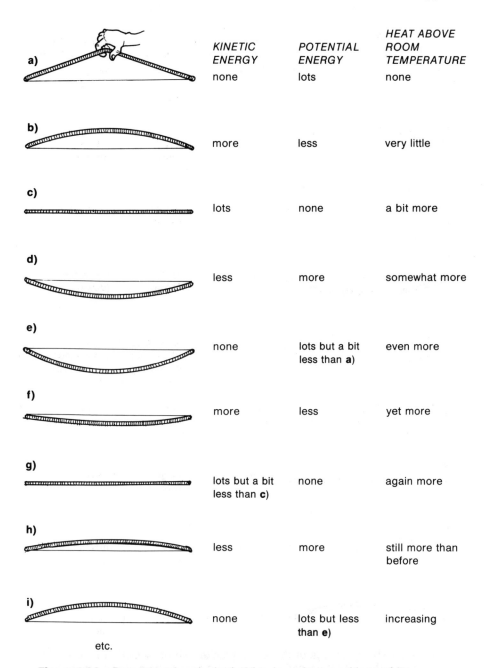

Figure 1.2A Drawings of a plucked string in various positions of its oscillation, together with the kinetic energy, potential energy, and dissipated heat at each stage of the oscillation.

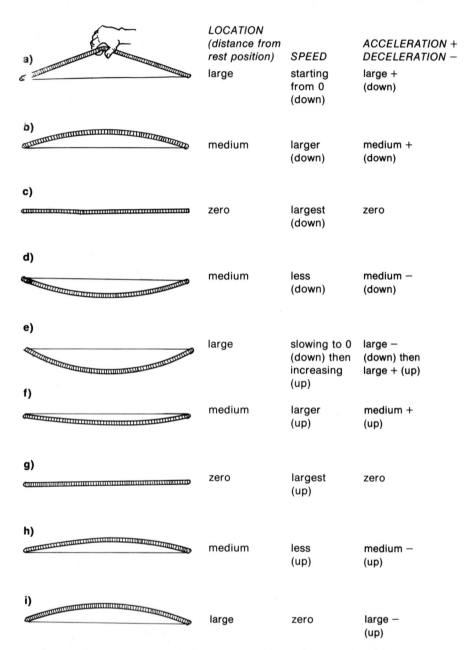

	LOCATION (distance from rest position)	SPEED	ACCELERATION + DECELERATION −
a)	large	starting from 0 (down)	large + (down)
b)	medium	larger (down)	medium + (down)
c)	zero	largest (down)	zero
d)	medium	less (down)	medium − (down)
e)	large	slowing to 0 (down) then increasing (up)	large − (down) then large + (up)
f)	medium	larger (up)	medium + (up)
g)	zero	largest (up)	zero
h)	medium	less (up)	medium − (up)
i)	large	zero	large − (up)

Figure 1.2B Drawings of the same plucked string shown in Figure 1.2A, together with the location, speed, and acceleration at each stage of the oscillation.

order to greater disorder. In science this rule is called "the second law of thermodynamics," and accounts for processes as diverse as a plucked string, a life cycle of a tiger, the evolution of galaxies, and the rise and fall of human civilizations. But for our purposes all we have to remember is that in any motion pertaining to sound and music there is always some heat loss, which eventually becomes predominant.

Summary

What we need to know about motion can be easily summarized and does not transcend the concepts, ideas, and experiences that arise in everyday life. We saw that motion can be described by providing a *table* in which we indicate the *location* of the object at all different *times*. To characterize the motion in a handier way, we can derive various indicators from this table. The *speed* tells the change in location per unit time. The *velocity* is speed combined with giving the direction in which the motion takes place. In particular, complete information on velocity can be given in terms of three *components* of velocity, which provide the speeds in the three dimensions in which we live. Speeds tend to change with time, and we can describe this change of speed through the concept of *acceleration (or deceleration)*, which gives the change of speed per unit time. The relationship between location and time, as a primary piece of information, and speed and acceleration (deceleration) as secondary, derived quantities suffice to give us a precise *description* of motion.

To get a glimpse of the *explanation* of motion also, we use two more concepts—force and energy. *Force* turns out to be proportional to the acceleration it causes, and also proportional to the mass (weight) of the object on which it acts. We can thus explain accelerations we observe by saying that a force is at play.

Energy in science comes in serveral forms. *Kinetic energy* (energy associated with large-scale motion) turns out to be proportional to the mass (weight) of the object that moves, and also proportional to the *square* of the speed with which the object moves, thus connecting energy with motion. *Potential energy* describes the latent capacity of an object to move, to acquire kinetic energy.

Heat is kinetic energy on a microscopic, molecular level. The universally valid law of the *conservation of energy* says that in a system isolated from the outside, the total sum of all forms of energy must remain the same at all times. Thus the time evolution of a system proceeds in the form of conversion from one form of energy to another, with the sum of the various forms of energy always the same. The conversion processes, on the average, prefer to move in the direction of converting *into heat*. This is the manifestation of another very general law (the second law of thermodynamics), which says that the universe evolves from a state of greater order into a state of greater disorder. This concludes our discussion of motion in general, and we can now turn to the discussion of the type of motion that prevails in phenomena of sound and music.

Chapter Two
Periodic Motion: Vibration and Oscillation

MUCH OF THE motion we will discuss in connection with sound and music will be of a special type in which something will move "back and forth." In other words, if you were to anchor yourself at a given point in space, you would see the "something" return to that spatial point from time to time, with regularity. The motion covers a sequence of locations, and in each of these a periodic return and departure can be observed.

Such periodic motion, also referred to as vibration or oscillation, can be seen in a large variety of happenings in the world. A person going to work in the morning and returning home in the evening can be said to oscillate between the workplace and home. The hands of a clock also describe a periodic motion. Migrating birds oscillate between their winter and summer habitats. The sun's movement across the sky is periodic. A rocking chair oscillates periodically. A weight suspended at the end of a spring and then distended will move back and forth periodically. A city bus making runs on a given line has an oscillating, periodic motion. The child on a swing oscillates back and forth. Certain electronic components are called oscillators, which implies that something in them oscillates. And the stock market average is also said to oscillate, although any regularity appears to be nonexistent and so we tend to describe its motion as "fluctuation," suggesting an irregular pattern of return to the same place rather than "oscilla-tion" or "vibration," which imply some regularity. We will use "os-cillation" and "vibration" interchangeably here. Strictly speaking, oscillation denotes the back-and-forth motion of anything, includ-ing the abstract (such as one's opinion oscillating between two ex-tremes), whereas vibration usually refers to the oscillation of a material, often rigid, object, such as a string. The physical essence is, however, the same, and for our purposes there is no need to dis-tinguish between the two.

Location–Time Table

To describe a periodic motion such as oscillation or vibration, we will use an example from everyday experience. Consider the city bus making its runs on a given line. We know that the bus's mo-tion can be completely described by a table that indicates its loca-tion at different times during the run. Table 2.1 is the location–time chart for our imaginary bus.

The table does not actually give the location of the bus at every point in time during its run, but only at those of interest to the passengers—when it is at the stations. However, that will not make any difference for our purposes.

Period and Frequency

By inspecting the table, we can easily determine some interest-ing characteristics of the motion of the bus. For example, we see that each run takes one hour and 20 minutes, and that the bus spends 20 minutes at the last stop, where it turns around. If, therefore, we look at the time elapsed between two departures of the bus from Fourth Street, heading toward 64th Street, this time is three hours, 20 minutes. Inasmuch as we view the bus's motion as periodic, we can say that the *period* of the motion is three hours and 20 minutes.

We can also see that this period can be determined not only by the passengers at Fourth Street, but also by passengers at any other stop. For example, the time elapsed between two consecutive

Table 2.1
Location–Time Chart

Station	Run #1	Run #2	Run #3	Run #4	Run #5
4th Street	2:00	5:00	5:20	8:20	8:40
10th Street	2:08	4:52	5:28	8:12	etc.
20th Street	2:20	4:40	5:40	8:00	etc.
28th Street	2:26	4:26	5:46	7:46	etc.
40th Street	2:40	4:14	6:00	7:34	etc.
48th Street	2:50	4:02	6:10	7:22	etc.
54th Street	3:02	3:50	6:22	7:10	etc.
64th Street	3:20	3:40	6:40	7:00	etc.

buses leaving 28th Street in the direction of 64th Street is also three hours and 20 minutes. Note that the time between consecutive passing buses as observed by a passenger at 28th Street is *not* 3 hours and 20 minutes, but alternates between two hours and one hour, 20 minutes. This is so because if one bus is seen to pass 28th Street going toward 64th Street, the next bus will be going in the opposite direction, the next after that again toward 64th Street, and so on. The time interval that describes the important regularity of the motion is the time between every *second* bus, namely, between those going *in the same direction.* You can ascertain this by noting that the time elapsed between every second bus is the same for every stop, whereas the time elapsed between consecutive buses varies greatly from stop to stop, depending on whether the stop is close to the terminus or not.

This characteristic of regularity is the period, which can also be expressed in another way. If you phone the bus company for information, you probably will not be told that the oscillatory motion of the bus has a period of three hours and 20 minutes, but that there are so many buses per hour. How can we calculate that? In our example there is one bus per three hours and 20 minutes, or $1/3\frac{1}{3}$ = 0.3 bus per hour (Figure 2.1 on page 26). This number is called the *frequency* of the bus run, to use the everyday meaning of a common word.

133, 459 Amplitude

So far we read from the table two strictly related characteristics of the bus runs—period and frequency—which should really be

LIBRARY
College of St. Francis
JOLIET, ILLINOIS

Figure 2.1 A literal illustration of a bus run with a frequency of ⅓ bus per hour.

considered one characteristic as one is simply the inverse of the other. These two quantities have to do with time. There is also another characteristic (which has to do with space) that can be determined from the table—the length of the line. It stretches from Fourth Street to 64th Street, which may be eight miles, let us say. This parameter of the motion is denoted by the word *amplitude,* from the everyday use of English in which ample refers to size, and so amplitude denotes the ampleness of the bus line. In scientific usage the amplitude is defined as the distance from the "middle" of the oscillatory motion to one of its extremes, so that in our case the amplitude would be half of eight miles, or four miles.

If we use this middle position as the reference point, we can say that most of the time the bus is "displaced" from this position, and we can characterize the position of the bus by giving its displacement (distance) from the middle point. Using this terminology we can then say that the amplitude is the largest displacement.

Thus we have charcterized the oscillatory motion by its period (or frequency) and its amplitude. As we shall see, these two parameters do not tell us everything about the oscillation, and some details will need to be further specified. But the two parameters do give us a rough idea of what the oscillation is like.

With these three characteristics to describe oscillations, we can return to the vibrating string shown in Figure 1.2A and B (see Chapter 1). The amplitude there is the largest displacement from the rest position, that is, the one shown in Figures 1.2a, 1.2e, or 1.2i. The period is the time elapsed between Figure 1.2a and Figure 1.2i, since after Figure 1.2i the process repeats itself. The number of times the motion cycle between Figures 1.2a and 1.2i repeats per second is given by the frequency. We can also understand that if the force pulling the string toward the equilibrium position is large, the acceleration will be large, therefore the string will move faster, resulting in a short period or a large frequency. This example ties together all the concepts we have studied thus far.

Plotting Oscillations

To see more, we should find a better way to visualize the motion of the bus than by looking at time tables. It is not a matter of getting more information, but of presenting the information in a handier way. The first step is to bring the designation of the stations closer to reality. In the city in which our bus runs, the blocks are extremely regular and equally spaced, so we can draw a map of the streets and use that as a distance scale. We then get a picture like that in Figure 2.2, in which the locations of the stations are depicted "realistically." The times, however, are still in a tabular form and so are not as easily visualized. We thus try yet

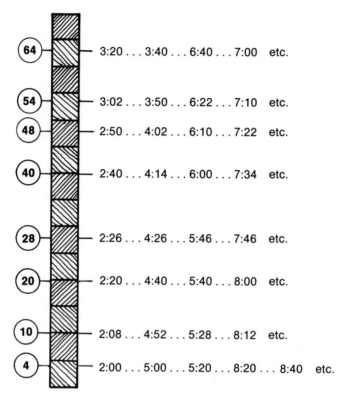

Figure 2.2 A timetable for a bus run with the distances (along the vertical scale) indicated in proportion (that is, stops farther apart from each other are shown at a greater spacing).

another picture (Figure 2.3) in which we depict the times in terms of appropriately proportional distances from the vertical line on which the stations are indicated. To do that we draw a scale line at the bottom of the picture, and each time point is placed above the appropriate value on the scale line. In a way this is a strange picture, because the *vertical* locations and distances on the picture mean actual distances on the city streets, whereas the *horizontal* distances mean time intervals. However, as humans are limited in their ways of perceiving things, any relationship between two quantities must be pictured in terms of two spatial dimensions, such as in illustrations in newspapers that portray the growth of the deficit of the federal budget with time. In such "graphs" the

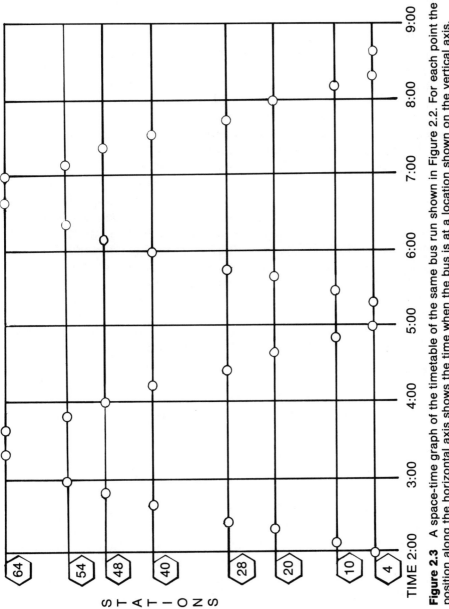

Figure 2.3 A space-time graph of the timetable of the same bus run shown in Figure 2.2. For each point the position along the horizontal axis shows the time when the bus is at a location shown on the vertical axis.

vertical direction represents dollars and the horizontal direction time, thus neither of the two spatial dimensions has anything to do with space.

As the human mind can deal with pictures much better than with tables in attempting to acquire an overview of a situation, from now on we will use such pictures to describe various aspects of periodic motions. Whenever you see such a picture, you should first establish what the two spatial dimensions mean. Two pictures may show very similar patterns and yet mean totally different things because the two dimensions (or axes) are labeled to represent completely different quantities.

Figure 2.2 or 2.3 (or, for that matter, the Table 2.1), by giving us the locations at various times, allows us to calculate the speed and the acceleration of the bus at various times. For example, on the first run (between times 2:00 and 3:20) the (average) speeds between stops are (in units of blocks per minute) 6/8, 10/12, 8/6, 12/14, 8/10, 6/12, 10/18, or decimally, 0.75, 0.83, 1.33, 0.85, 0.80, 0.50, and 0.56. Therefore, the (average) accelerations (or decelerations) between these seven intervals will be, in units of blocks per minute squared and using plus for acceleration and minus for deceleration, $(0.83 - 0.75)/(2{:}14 - 2.04) = +0.008$ (where the halfway time for each interval was used to calculate this average acceleration) for the first two intervals; +0.056 between the second and the third; −0.048 between the third and the fourth; −0.004 between the fourth and fifth; −0.027 between the fifth and the sixth; and, finally, +0.004 between the sixth and the seventh. Make sure you can reproduce these numbers by your own calculations.

Shapes of Oscillations

What Figure 2.3 tells us easily (in addition to the period and amplitude) is the "shape" of the oscillation. Much can be deduced about the details of the bus run by looking at details of this shape, as some of the chapter exercise problems at the end of the book show.

If, instead of looking at the bus runs as oscillations, we similarly investigate, for example, the oscillatory motion of a weight at the end of a spring, we get a differently shaped oscillatory pattern.

It will be much closer to the shape shown in Figure 2.4b. It rises fast from its middle point, turns over, gently hits the maximum, gently turns around, and then drops more and more rapidly, passes through the middle point rapidly, and again turns over to hit the other extreme point gently, and so on. Such curves (or curves very close to them) are extremely common in the various oscillations we can observe in nature. In mathematics they are called "sine curves." Mathematicians, of course, have a very precise quantitative description of sine curves, but we need not be concerned with that in our attempt to understand the physics of music.

By no means are all oscillation curves sine curves, or even close to them, but since sine curves constitute such a good approximation of many physical systems, they are chosen to describe even the more complicated curves. How this is done is the next item on our agenda. The first step is to explain what we mean by "a superposition of oscillations."

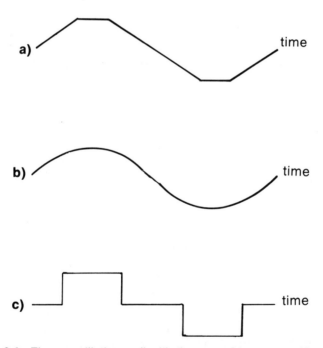

Figure 2.4 Three oscillations, all with the same frequency, with quite different curve shapes. **(a)** This is close to the curve for the bus run described in Figures 2.2 and 2.3. **(b)** A "sine" curve. **(c)** A "square" curve.

"Adding" Oscillations

To illustrate superposition we consider another example of an oscillatory motion, that of a mailcarrier delivering mail along a street and alternating back and forth between the two sides of the street (Figure 2.5). The mailboxes along the street are arranged equidistantly and in a staggered array as shown. It takes the mailcarrier one minute to walk between two consecutive mailboxes and put mail into one of them. This work, therefore, can be described as an oscillatory motion with a period of two minutes, or a frequency of one-half delivery on a given side of the street per minute. If the width of the street is 100 feet, the amplitude of the oscillations between mailboxes is 50 feet.

Our mailcarrier has a favorite dog that needs some exercise, and so one day the dog is taken along on the mail route, on a 10-foot leash. The dog is very active and has an almost neurotic habit of running back and forth across the path of the mailcarrier as far as the leash will allow. It takes 20 seconds for the dog to run from the extreme left of the mailcarrier to the extreme right.

When our mailcarrier is at a standstill, the motion of the dog can be easily described as an oscillation with a period of 40 seconds, or a frequency of 1½ per minute. The amplitude of this oscillation is the length of the least, or in this case 10 feet.

Now we want to describe the motion of the dog when accompanying the mailcarrier on the mail route. The motion will obviously be composed somehow of two different oscillations: that of the mailcarrier about the centerline of the street, and that of the dog around the mailcarrier. Indeed, the way we do this composition is obvious. We first plot the path of the mailcarrier, and then indicate the usual pattern of the dog's oscillation, but *always measured from where the mailcarrier is at that particular moment*. In other words, the composite oscillation the dog will have can be obtained by adding the relative displacement of the dog from the owner to the mailcarrier's displacement from the center of the street. This is shown in Figure 2.5a. In some cases the two displacements are in the same direction, and hence literally add, but at other moments the mailcarrier may be to the right of the center of the street and the dog may be to the left of the mailcarrier, and hence the two displacements are really subtracted from each other. We might say that we add the two displacements, but take into account their directions.

Nothing Out of Something

Consider the following interesting situation that may arise from such a superposition of two oscillations. Let us change the situation of our dog-owning mailcarrier by substituting a more sedate dog, which is allowed on a longer leash. In particular, let us assume that the leash is now 50 feet long, and the dog ambles over from an extreme right position to an extreme left one in one minute (Figures 2.5b and 2.5c). In that case, if the mailcarrier is clever enough to figure out the proper time to start to walk the route as compared with the habits of the dog, it can be arranged that although the mailcarrier oscillates with respect to the street, and the dog with respect to the mailcarrier, the resulting path of the dog will be a straight line along the center of the street—in other words, seemingly no oscillation at all (Figure 2.5b). All the mailcarrier has to do to achieve this is to cross the centerline of the street, and go to the right, just when the dog crosses over in front, going to the left. Since in the case of this second, more dignified dog the amplitudes and the frequencies of the oscillations of the mailcarrier around the centerline of the street and of the dog around the mailcarrier are the same, such a complete cancellation of the two oscillations becomes possible, provided that when the mailcarrier starts deviating from the centerline to the right, the dog starts deviating from the centerline to the left. In other words, the two oscillations are just half a period apart. In scientific terminology being half a period apart is often expressed in a different way. The entire oscillation is assigned 360 degrees in that notation, with 0 degrees where the oscillation starts, say, from the centerline to the right. Then the *phase* of the oscillation at a given time is defined as the number of degrees corresponding to that particular displacement in the picture of the oscillation (Figure 2.6). Thus, to return to our mailcarrier and the dog, if the phase *difference* of the oscillations of the mailcarrier and of the dog is 180 degrees, we can get a complete cancellation of the two oscillations; that is, the *resultant* of the two oscillations (the oscillatory pattern resulting from the superposition of the two oscillations) is no oscillation at all. This phenomenon is also called *complete destructive interference* of the two oscillations. To help you visualize how this comes about, an intermediate case of the more sedate dog on a 40-foot leash is described in Figure 2.5d.

If the phase difference of the two oscillations is 0 degrees,

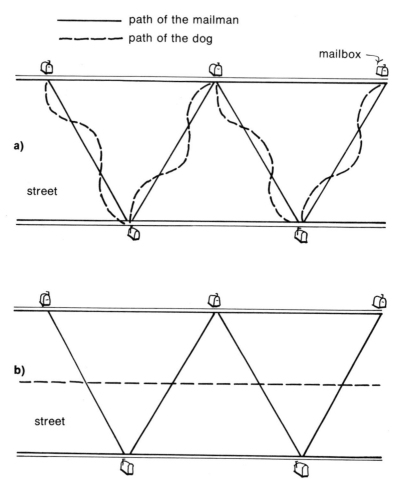

Figure 2.5 Locations of the mailman (thin solid line) and the dog on a leash (broken line), under various conditions regarding the length of the leash and the frequency of the dog's oscillation relative to the frequency of the mailman's alternation between the two sides of the street (heavy solid lines).

(a) Period of the mailman's, two minutes; period of the dog, 40 seconds; length of the leash, 10 feet.

(b) Period of the mailcarrier, two minutes; period of the dog, two minutes; length of the leash, 50 feet; phase difference between the mailman and the dog, 180 degrees.

(c) Period of the mailcarrier, two minutes; period of the dog, two minutes; length of the leash, 50 feet; phase difference between the mailcarrier and the dog, 0 degrees.

(d) Period of the mailcarrier, two minutes; period of the dog, two minutes; length of the leash, 40 feet; phase difference between the mailcarrier and the dog, 180°.

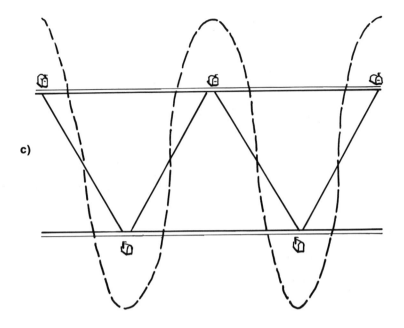

c)

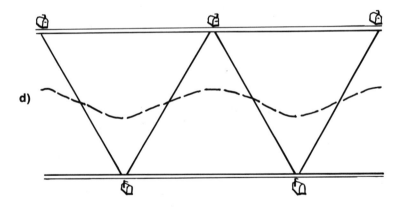

d)

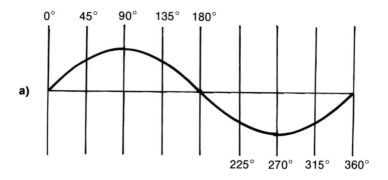

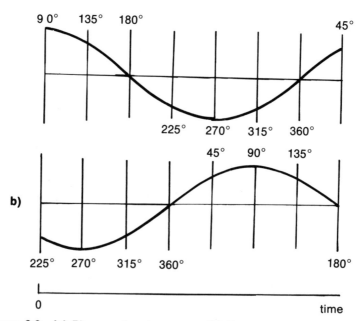

Figure 2.6 **(a)** Phases of a sine curve. **(b)** Two curves with a phase difference of 225 degrees (or −135 degrees). The upper curve is "ahead" of the lower one by 225 degrees (or, if you prefer, "behind" the lower one by 135 degrees).

however, our dog will describe huge oscillations with a 100-foot amplitude (Figure 2.5c). In this case the two oscillations reinforce each other, and we have a completely *constructive* interference. But the way to construct the location of the dog at any given moment remains the same: we simply add to the mailcarrier's position the position of the dog with respect to the mailcarrier. Note that we could also first take the dog's location in the dog's oscillations around a stationary mailcarrier, and add to that the oscillations of the mailcarrier. In other words, the order in which we do the addition makes no difference.

More About Adding Oscillations

Now that we understand the superposition of oscillations, let us return to the analysis of strange oscillation shapes in terms of sine curves. As mentioned, since sine curves are so common, we will try to analyze other shapes in terms of sine curves as well. Is it perhaps possible to describe such a strange shape as a superposition of such sine curves?

The answer is a resounding "yes," thus establishing an extremely powerful way of describing any oscillation shape. How this can be done is explained in the following.

We consider an oscillation curve with a definite period and a completely arbitrary shape. We mark the beginning of the period where the curve is zero. To describe this oscillation, we use sine curves with yet unspecified amplitudes but with periods that are two times, 2/2 times, 2/3 times, 2/4 times, 2/5 times, and so on, the period of our strangely shaped curve to be described, and which all go through zero at the beginning and end of their periods. These curves are illustrated in Figure 2.7. We then have an *infinite* number of curves, the periods of which have the relationships listed. Furthermore, we can take these curves so that at the left-hand end point of the curve to be analyzed, all curves are "in phase," that is, they have 0-degree phase difference. This is also shown in Figure 2.7 on page 39.

The game then is to assign the *amplitudes* of this infinite set of curves so that when they are superimposed on top of each other, the resulting curve *is* the particular curve to be analyzed. One might first think that since there is an infinite number of different curves to play with, success is assured. But then consider that this superposition should be possible for any curve to be analyzed, and although for each strange curve the amplitudes can be readjusted anew, these curves may have so many crazy little wiggles in them that even with an infinite number of curves, each of which is smooth, we may not be able to achieve the goal. So the outcome of the attempt may be in doubt.

In actuality, however, it is not. The brilliant French mathematician Fourier, more than 150 years ago, proved that such a superposition into sine curves of the set we specified can be done for any curve that can arise in physical situations. The analysis of a curve of arbitrary shape into such sine curves thus is called the Fourier decomposition of a curve. A "recipe" can be given for any curve, that is, a series of numbers that denote the amplitude of each sine curve such that when one superposes all those sine waves with those specified amplitudes the original curve is obtained.

If the sine curves are numbered according to twice the ratio of the original curve's period to the sine curve's period (indicated on the right hand margin of Figure 2.7) and that number is used to label the horizontal axis of a graph, while the vertical axis is labeled by the size of the amplitude of that sine curve, the result is as shown in Figure 2.8 on page 40. Each picture displays the recipe for composing a given oscillation curve out of the set of sine curves. Some examples of such a superposition of curves out of sine waves are given in the chapter exercises.

Representing Kinks

There are some qualitative conclusions one can draw about how these superpositions function even if one does not have Professor Fourier's brilliance and quantitative mathematical facility. Consider, for example, a curve with a sharp "kink" in it, that is, with a rapid variation somewhere along it. It sounds intuitively obvious that with the use of only the first few sine curves on our list (i.e., only the lowest Fourier components), it will be difficult to produce such a kink even in a superposition, since these first few sine curves change with time in a rather gentle and leisurely way, quite

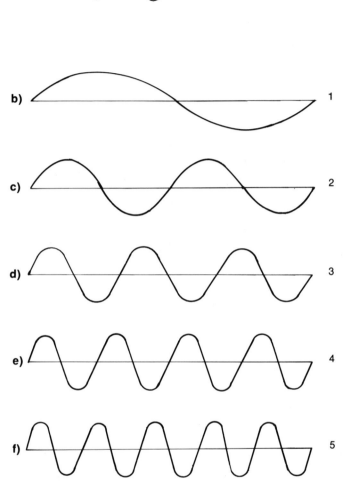

Figure 2.7 **(a)** An arbitrary curve with a "wild" shape. The curve repeats it-self after describing the oscillation indicated on the figure.

(b)–(f) A few of the simplest sine waves needed for the Fourier decom-position of the waves shown in (a). These sine curves are shown here with equal amplitudes. In the recipe for the decomposition of (a) into these waves, the amplitudes of the various sine curves are specified and depend on the shape of curve (a). (See Figure 2.8).

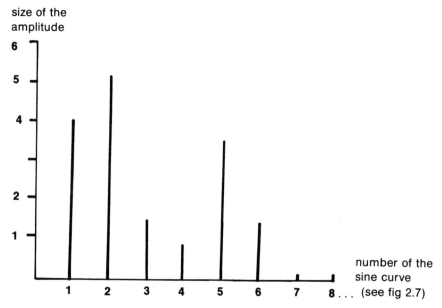

Figure 2.8   The pictorial specification of the amplitudes in a Fourier decomposition. The lengths of the vertical lines indicate the amplitudes of the various sine waves shown in Figure 2.7. This is called a spectrum.

in contrast to the kink. We will, therefore, need high Fourier components (that is, rapidly oscillating sine curves) to produce the kink. Furthermore, only one such high Fourier component will not do, since it would give such kinks all over unless destructively interfered with almost everywhere by Fourier components of roughly equally frequent oscillations. Thus rapidly varying oscillatory curves can be reproduced only if one has a good assortment of high Fourier components (sine curves with very large frequencies) at one's disposal. This feature will arise very frequently in our discussions of sound and music, as, for example, in discussing the "high-fidelity" requirements for sound reproduction.

Summary

In the discussion of sound and music, we often encounter back-and-forth motion around a fixed point in space. Such motion is called vibration or oscillation. We can describe this motion by providing its time table. The time needed to describe one whole round of such a motion is called a period, and its inverse, which tells how many times per unit time the motion traverses the same stage, is called frequency. The size of the deviation in space from the average position is called the amplitude of the oscillation. Oscillations can be symbolically depicted by a graph on which the vertical direction measures the spatial deviation from the average and the horizontal axis gives the time. Oscillations can be added (or "superimposed") by adding, at each moment in time, the two deviations of the two oscillations. This recipe can lead, in some cases, to no deviation at all for the resulting oscillation at any time, namely, when the two oscillations to be added have the same frequency and amplitude but one moves one way when the other moves the other way. A certain smoothly shaped oscillation curve is called a sine curve. It can be shown that any periodic oscillation curve, no matter how strange in shape, can be represented as a superposition of an infinite number of sine curves with periods equal to two times, 2/2 times, 2/3 times, and so on, the period of the initial curve. The resulting set of amplitudes is called the Fourier spectrum of the strangely shaped curve. Curves with many kinks—that is, with sudden changes in them—need a goodly number of high-frequency Fourier components to describe them.

This completes our discussion of what we need to know about oscillations in general. In the next chapter, we use this knowledge in the discussion of another concept that is very important to understanding sound and music—the discussion of waves.

Chapter Three
Waves

I N DEALING WITH waves, you can again rely on your personal experience to visualize and work with the concepts we will discuss. All of us know about waves, perhaps most directly through waves in water. But we also hear about seismic waves propagating in the inside of the earth during an earthquake, or shock waves traveling in the air during a sonic boom. We also talk about waves of excitement traveling through a crowd, or waves of rumor spreading among the population, or a wave of epidemic hitting a community.

Traveling Waves

As it turns out, we need not stretch our imagination much to understand what a wave is, once we have learned about oscillations. In the case of an oscillation, there was one object, at a given point in space, which left that point to go back and forth around this former equilibrium position. The wave is simply a whole array of neighboring objects oscillating, stretched out along a line or a surface. For example, in the ocean, every part of the surface water oscillates up and down. Originally only some part of the ocean was whipped up by the wind or a passing boat to oscillate, but since neighboring water molecules on the surface of the ocean are tenuously but definitely linked with the one that started oscillating, these other molecules are also dragged along, and also will oscillate. Because of their sluggish response, however, their oscillation up and down will be a bit delayed compared with the oscillation of the originally whipped-up molecule, and the next molecules will be even more delayed. In other words, the various molecules will all oscillate, but with an increasingly larger phase difference as compared with the initially oscillating molecule as we move farther and farther from the latter. The *appearance* of this ensemble of staggered oscillations will give the impression that the top of the ocean is *traveling away* from the initial oscillators (like the water waves appearing to travel away from the location where one drops a stone into the water), even though each water molecule stays almost entirely in its original location and merely bobs up and down. We call such a phenomenon a *traveling wave.* You should keep in mind, however, that even though the protruding part of the water (which we call a wave) does travel, the water of which the wave is a part does not, except for bobbing up and down. You can check this by floating in the ocean (far enough in so you avoid the "breakers") and observing that both the water around you and yourself move mainly up and down while the waves pass at considerable speed. Such a traveling wave is shown in Figure 3.1, which we discuss in more detail shortly.

Standing Waves

Not all waves travel, however. We can also have a situation when an ensemble of neighboring objects all oscillate, but in *unison.* In fact, we already encountered, in a different context, such a system in Figure 1.2A and B, where we had a stretched string that was plucked and then released. The various parts of that string then proceeded to oscillate up and down, but since in a *solid* object (like the material of the string) the neighboring molecules are connected more tightly, and also because the ends of the string were tied down, these oscillations were all in phase. Such a wave, therefore, is called a *standing wave,* since the appearance of this en-

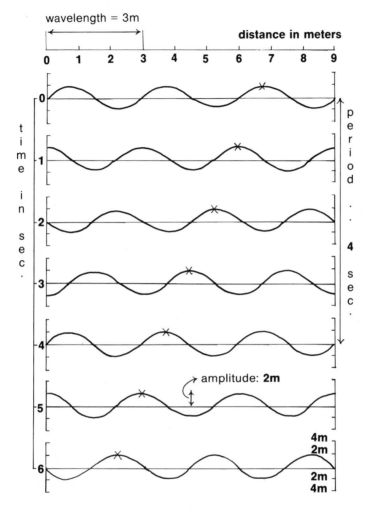

Figure 3.1 A sequence of a transverse traveling wave at different times. The distance scale is indicated at the top along a horizontal line, and the time when the snapshot was taken is shown along a vertical line on the left-hand margin. The x marks a particular crest at various times, to illustrate that although each point just oscillates up and down and remains in the same horizontal position, the wave *pattern* moves.

semble of oscillations gives the impression that the most protruding part of the string is at a standstill (except for bobbing up
and down).

Thus we see that in both traveling waves and standing waves
there is a *collection* of point objects (e.g., the surface water
molecules in an ocean wave) that oscillate. The difference is that in
the standing waves all the oscillations have the same phase (i.e.,
they are all "in phase"), while in a traveling wave the oscillations of
neighboring points are staggered, and as you move in space, the
oscillations gradually fall more and more behind in phase.

Transverse and Longitudinal Waves

In the foregoing examples, the waves were stretching along a
direction that was perpendicular to the direction along which the
bobbing up and down occurred. Because of this such waves are
called *transverse* waves. In these waves the oscillation propagates
to the neighboring location because of links between neighboring
molecules. Such links exist in solids and liquids, but not in gases.
Hence such transverse waves can be produced only in solids or
liquids.

We can also have waves in gases, but of a slightly different kind,
as shown in Figure 3.2. Consider a set of gas molecules lined up
along a straight line, as in Figure 3.2a. If we push the left-most
molecule to the right (Figure 3.2b), it will eventually collide (Figure
3.2c) with the next molecule to the right, push it to the right, and
bounce back to the left. This next molecule (Figure 3.2d), in turn,
will bounce against the next one on *its* right, push it to the right,
and itself bounce back. This chain of collisions and bouncing back
continues, and repeats as the molecules on the far left bounce back
to the right again (due to collisions with molecules or with a wall
on their left). Again, as in the case of the transverse wave, each
molecule stays approximately in the same location, except for oscillations around that location to the right and the left. The *appearance* of the ensemble of right–left oscillations, however, gives
the impression of a high-density bunch of molecules traveling
from the left to the right, followed by a low-density bunch of
molecules, and so forth. Just as in the case of the transverse waves,

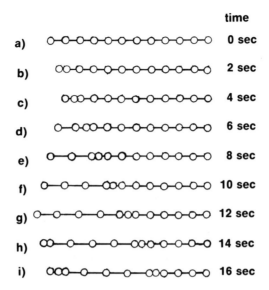

Figure 3.2 A sequence of a longitudinal traveling wave, similar to the snapshots shown in Figure 3.1 for a transverse traveling wave. The condensation of points form a pattern that travels from left to right as time goes on.

where only the protrusion at the top of the water seems to travel and each water molecule actually stays where it is except for bobbing up and down, here also only the condensations and rarifications of the molecule *patterns* appear to travel, while the individual molecules stay put except for oscillations right and left. This kind of a pattern is called a traveling *longitudinal* wave, because the individual molecular oscillations are in the *same direction* as the extension of the wave.

Similarly, we can also produce standing longitudinal waves, as shown in Figure 3.3. Here, some molecules do not move at all from their original positions, some oscillate a little right and left, some oscillate considerably right and left (just as in the standing transverse wave) and some molecules in the wave bob up and down a great deal, others less so, and some (namely, the two end points in Figures 1.2A and B) stay put. Those points on a standing wave (longitudinal or transverse) that *always* stay put are called *nodes*.

As mentioned, in gases we have only longitudinal waves—but in liquids and solids, we can have both transverse and lon-

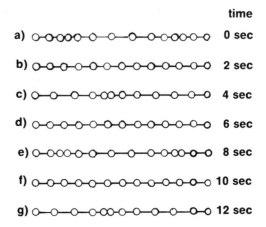

Figure 3.3   A sequence of a longitudinal standing wave. Note that at certain positions along the horizontal line, the distance between the dots never changes. These are the nodes of the longitudinal standing wave.

gitudinal waves. Although only longitudinal waves are possible in air, we actually *draw* all waves (whether in air, in a violin string, or elsewhere) *as if* they were transverse waves, simply for convenience, since drawing the many little dots—as in Figures 3.2 and 3.3—is tedious and difficult to visualize. You should remember, however, that in reality the waves in the air are always longitudinal.

Characterizing Waves: Frequency and Wavelength

Since waves are just a set of oscillations, we can carry over our characterization of oscillations to waves and use the period (or frequency) to describe the time development of these oscillations, and the amplitude to describe the largest size of the deviation from the equilibrium position. For waves, however, we need an additional quantity to describe the variation of the wave *in space*. Since the oscillations of the individual points in the wave are periodic in *time*, the variation of the wave in *space* will also be repeating, in that after a certain distance in space, the wave will repeat itself. The distance of this periodicity of the wave in space is called the *wavelength*, a natural term indeed, considering that it tells you how long the wave is in space. In Figure 3.1 the wavelength of the wave is indicated.

Figure 3.1 offers a sequence of snapshots of the wave. Each snapshot covers the same nine meters of space, and the snapshots are taken one second apart. Such a set of pictures gives a complete picture of the wave, since one can read from it both the spatial extent and shape of the wave, as well as its time development. This time development could also be pictured differently—with us anchored in a given location in the wave and, with stopwatch in hand, measuring the time variation of our bobbing up and down at that location as the waves coming by toss us to the crest and, a bit later, into the trough. If we then plot this time variation in our usual way (see Chapter 2), namely, using the horizontal direction to represent time and the vertical direction to represent the displacment from the equilibrium position, we obtain Figure 3.4. Make sure that you completely convince yourself that the time variation of the wave shown in Figure 3.1 and the time variation of the wave shown in Figure 3.4 are the same but represented in different ways. If you understand the connection between Figures 3.1 and 3.4, you are likely to understand waves sufficiently for our purposes. For example, can you tell in what location on Figure 3.1 I was anchored when I prepared Figure 3.4? If you can answer this question, you have understood much of what we said about waves and oscillations.

Relating Frequency and Wavelength by the Speed of Propogation

As mentioned earlier the spatial extent of the wave is repetitious *because* the oscillations of the individual points are periodic. This implies, of course, that the spatial and time characteristics of a wave are not independent of each other. Indeed, the period (or frequency) and the wavelength are connected to each other in a simple way if you consider a traveling wave. Imagine yourself again in the ocean, anchored at a given point and bobbing up and down

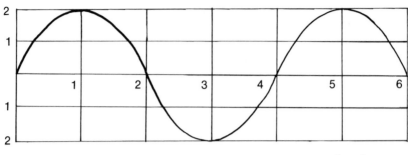

time in sec.s

Figure 3.4 A time plot of the wave of which the space plot is shown in Figure 3.1.

with the waves. Let us say that when you are at a given crest, you start your stopwatch. As time passes you are let down into a trough, and then lifted up onto the next crest. At that point you stop your stopwatch and thus determine the period of the oscillation. At the same time, you look in the direction in which the waves travel, and see the *previous* crest (on top of which you started your stopwatch) a certain distance away. That distance is, of course, one wavelength, since the definition of the wavelength can be given as the distance between neighboring crests. Thus you can determine that during one period of the oscillation, the wave traveled a distance of one wavelength. The *speed of propagation* of the wave, therefore, is the wavelength divided by the period (i.e., the distance divided by the time interval). Since the frequency is the reciprocal of the period, one can also say that the speed of propagation of a traveling wave is equal to the wavelength times the frequency. This relationship will come in handy later when we discuss the musical notes produced by instruments of various sizes.

The foregoing argument apparently does not hold for standing waves as they do not have a speed of propagation. Since, however, it can be shown (but we will not do so) that standing waves of a given wavelength can be composed of the interference of two traveling waves of the same wavelengths, the relationship among wavelength, frequency, and the speed of wave propagation in the medium in which the standing wave is formed also holds for standing waves.

Damped Waves

Since waves are just a sequence of oscillations, the superposition of oscillations discussed in the previous chapter naturally also holds for waves. Indeed, the various interference phenomena we discussed for oscillations also hold for waves.

As waves travel in a gaseous medium, some of the energy contained in the oscillations of the gas molecules transforms into disordered heat motion of these molecules, and hence the energy contained in the ordered oscillatory motion (and in the wave motion) is reduced (since the total energy must be conserved). The *energy* of the oscillation is connected with the *amplitude* of the oscillation: the larger the amplitude, the more energetic is the oscillation. (This makes intuitive sense, doesn't it?) Thus the dissipation of energy into heat and the consequent energy loss in the oscillation itself will decrease the amplitude of the oscillation, and so of the wave. This is called *damping* of the oscillation or of the wave. A damped wave is shown in Figure 3.5. Can you tell in what direction the wave is traveling?

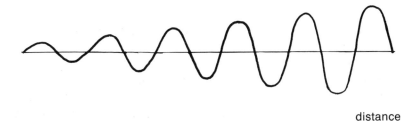

distance

Figure 3.5 A damped wave.

Waves at a Boundary

When a traveling wave hits a perpendicular "wall," that is, comes to a boundary of two different media, some of it will penetrate the wall (or the other medium). (You may think this strange, since in the case of ocean waves hitting a stone wall, the penetration is practically zero. Sound waves hitting a wall, however, partly penetrate, otherwise you could not eavesdrop through a closed door.) Some other portion of the wave will turn and come back into the first medium, going in the opposite direction. This process is called *reflection*. The more different the two media are, the more of the wave will reflect and the less will go through into the other medium. You might say that the wave motion is disturbed by the sudden, large change in the properties of the elements in the medium that oscillate, and that the larger this shock is, the more difficult it is for the wave to continue into the other medium and the easier it is to turn around and travel in the familiar first medium.

The reflected wave will have the same period and wavelength as the original wave, but its phase will depend on the comparative properties of the first and second media. If the second medium is less dense than the first, the reflected wave will have zero phase difference with the incoming one. If, on the other hand, the second medium is denser than the first, there will be a 180-degree phase difference between the incoming and reflected waves. It is as if the denser second medium "grabbed" the incoming wave and turned it around in phase, whereas if the second medium is less dense, it has not strength to do so.

If the wave hits the boundary, not from a direction perpendicular to the boundary, but at some angle, we can think of the wave as having two components (see Chapter 1), one perpendicular to the boundary and the other parallel to it. The latter, of course, will travel unimpeded. The first, on the other hand, will be partly transmitted into the second medium and partly reflected in the way discussed above. If we then combine the unimpeded part of the wave, which is parallel to the boundary, with the reflected part of the wave, which is perpendicular to the boundary, we get a wave that is now traveling away from the boundary, at an angle that is the same as the angle of the incident wave.

All these transmission and reflection processes are illustrated in Figure 3.6. In Figure 3.6a a wave perpendicularly hits a boundary on the other side of which there is a rarer medium. The reflected wave has the same wavelength, a slightly smaller amplitude, and the same phase as the incoming wave. On the right-hand side, a small-amplitude transmitted wave is seen. Figure 3.6b shows the same reflection process for the case when the medium on the right is denser than the one on the left. The reflected wave in this case again has the same wavelength and a smaller amplitude, but

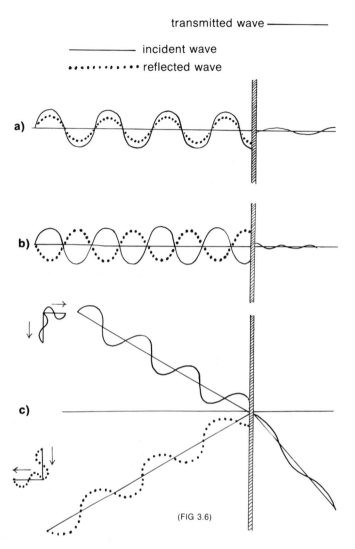

Figure 3.6 Reflection and refraction of a wave when impinging on a boundary of two different media.

(a) Denser medium on the left, rarer medium on the right. Note that the reflected wave (dashed line) is in the same phase as the incident wave, but has a smaller amplitude. The transmitted wave has a small amplitude and a long wavelength.

(b) Rarer medium on the left, denser medium on the right. The reflected wave has a 180 degree phase difference with the incident one. The wavelength of the transmitted wave is smaller.

(c) Reflection at an angle. The reflected wave makes the same angle with the wall as the incident wave, but the transmitted wave becomes "refracted" and makes an angle with the wall that is different from that of the incident (or reflected) wave.

shows a 180-degree phase difference with the incoming wave. Figure 3.6c shows reflection at an angle. (The right-hand medium is less dense than the left-hand medium.) Two other characteristics are noteworthy. The transmitted wave traveling in the other medium has a wavelength different from that of the incident (or reflected) wave, since in general the speed of propagation in different media is different, but the frequency of the oscillations remains the same as the wave passes the boundary. This can be understood by realizing that the molecules in the first medium at the boundary oscillate and pull with them the molecules in the other medium at the boundary, which therefore are forced to oscillate with the same period as the molecules in the first medium.

Note also that the *direction* of the propagation of the transmitted wave in the second medium in Figure 3.6c is different from the direction of propagation of the incident wave in the first medium. This phenomenon, which is also caused by the difference in the speeds of propagations in the two media, is called *refraction.*

The Waves We Know

Before we turn to the last phenomenon connected with waves that we need for our study of sound and music, let me give a quick overview of the kind of waves we commonly encounter in everyday life. The sound waves that we can hear directly with our human hearing mechanism are usually vibrations in air. Their frequency range is 20–20,000 per second, and since the speed of propagation of sound in air is about 1000 feet per second, the corresponding wavelengths are from 50 feet (corresponding to the frequency of 20 per second) to about a half an inch (corresponding to a frequency of 20,000 per second). Actually frequencies above 3000–4000 per second play only an auxiliary (though important) role in everyday sound phenomena, so the lower end of the wavelength scale is, in that respect, more at three to four inches. You should familiarize yourself with these rough estimates since they have an important role in explaining the qualitative behavior of sound in many situations.

Another type of wave very commonly encountered in everyday life is called an "electromagnetic" wave. This is a wave in which the strength of electricity and magnetism varies in a wavelike way. A very broad range of apparently unrelated phenomena fall under the broad umbrella of electromagnetic waves: radio and television waves, radar, radiation of heat, the colored light we see with our eyes, x-rays, "gamma" particles in nuclear and elementary particle phenomena, and so on. These various phenomena differ from each other only in frequency, and hence wavelength. In particular, the wavelength of your favorite FM station is about 10 feet, of your AM radio station around 1200 feet, of the light you see about 1/50,000 inch, and so forth. These are again rough magnitudes that will help you to understand differences in behavior among these various kinds of electromagnetic waves.

There is one crucial difference between sound waves and electromagnetic waves. As said earlier, sound waves are vibrations of air molecules, or molecules of other substances in which sound can propagate, such as water, steel, helium gas, or bones. Whatever the substance, sound waves need such a material medium in which to propagate.

In sharp contrast, electromagnetic waves are waves of the strength of the electricity and magnetism at various locations in space, and hence do not need a medium for their existence; it is not a material particle that vibrates in that wave, but only the amount of electric and magnetic "field," which can exist disembodied from material. Therefore, an electomagnetic wave can also propagate in vacuum, in empty space, since it is not something material that oscillates. We know this well and can ascertain it easily, since we can see the sun and feel its heat radiation, even though most of the space between the earth and the sun is very high vacuum, that is, practically devoid of any material particles. Anything happening on the sun can be seen but not heard.

In our studies of sound and music, we deal mainly with sound waves, but also involve electromagnetic waves when we discuss the transmission and storing of sound and music in Chapter 14.

Diffraction

Now that we have an idea of the various kinds of waves, and of the approximate size of their wavelengths and frequencies, we can turn to the last of the wave phenomena needed for our studies, that of *dif*fraction (as distinct from *re*fraction mentioned earlier).

Let us first recall some observations that we will relate to this so-far-unspecified "diffraction" process. When a light source encounters a nontransparent obstacle of the size we usually have around us (above an inch or so), that obstacle throws a sharp shadow (Figure 3.7a). If a source of sound is located around the corner of a building, one can still hear that sound (Figure 3.7b). When an AM radio wave is obstructed by a house, a hill, or some other object of similar size, one can still receive the station. In contrast, you have a much greater difficulty receiving an FM station when there is such an obstruction. A post driven into the ocean floor does not throw a "shadow" with respect to the large ocean waves rolling in. The waves seem to pass through it and the wave patterns are in no way different behind the post than before it. On the other hand, a 50-food wall built into the ocean floor does protect the ocean surface behind it, and the ocean waves rolling in from the outside can pass by at the ends of the wall but cannot penetrate the area behind it (Figure 3.7c).

To understand all this, let us take the example of a sound source around the corner and analyze how its waves propagate. To do this, let us first consider Figure 3.8, which illustrates how we can construct the propagation of a wave. Let us assume that we have a set of points (along the solid line in Figure 3.8) at which the oscillations of the various points have zero phase difference. That means that at all these points a crest occurs at the same time. If we want to determine how the oscillations are transmitted to neighboring points, that is, how the wave propagates, we need only construct the wave going out of each of these points and then take the superposition of all these waves. This is shown in Figure 3.8 where, a certain time later, the oscillation of each point has been transmitted to all points that lie on a circle around the original point, the radius of the circle being determined by the distance the wave can travel in that time period. By constructing these circles, we get a pattern like that shown in Figure 3.8. In fact in that figure only a certain number of points along the line are indicated with their circles around them, so as not to crowd the figure. The figure is a circle because we assume that the disturbance travels at the same speed in all directions.

It can be shown that if we superpose all these circular waves coming out of each point of the line, we find that their constructive and destructive interference at various places results in the points lying on the dotted line oscillating without any phase difference among themselves. There will also be oscillation at other points in space, but those oscillations will have a phase difference as com-

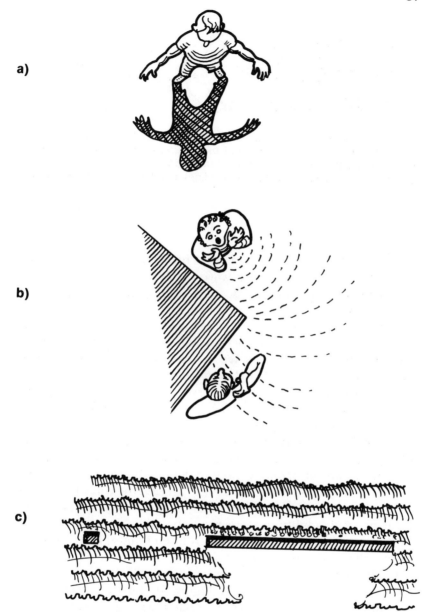

Figure 3.7 **(a)** Light casts a sharp shadow. The wavelength of light is tiny compared with our everyday distance scales, hence we see the shadow as "sharp."

(b) Sound does not case a sharp shadow: we can hear around the corner. The wavelength of sound is comparable to or larger than our everyday distance scale of a foot or so, hence sound "diffracts" around the corner to an extent that is quite appreciable to us.

(c) The ocean wave "passes through" a small pile but a long pier casts a shadow. The wavelength of the ocean wave is large compared with the size of the pile, but small compared with the size of the pier.

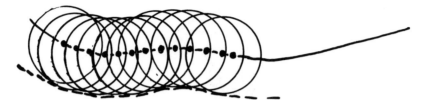

Figure 3.8 Constructing the propagation of waves. Each point on the solid line generates a circular wave, and the interference of the many circular waves creates the overall wave pattern (dotted line) at a somewhat later time. Thus we see the line of the wave (solid line) a bit later as another wave line (the dotted line).

pared with those on the dotted line. Thus, looking at the ensemble of oscillations together, as a wave, we find that the wave crest moves, during the time in question, from along the solid line to along the dotted line.

Now let us look at Figure 3.9. Originally the oscillation starts at point 0, and spreads, in the way explained on the previous figure, to increasingly larger circles. We also see that the wave can spread even to areas that are not "in line of sight" to 0, for example, to the area "around the corner" as compared with the location of 0. This is so because the oscillations at a point such as *F* are a result of the circular waves coming from points *D, E, C,* and many, many others, and there is no reason why those waves cannot reach an area that is "around the corner" from the original point zero. As it turns out (something we cannot show here in detail), the total constructive interference of all the waves reaching "around the corner" is not as strong as in places that are in "line of sight" of zero, and so the resulting total amplitude of the wave should be decreased around the corner compared with places in line of sight, but still some waves will reach around the corner.

If, on the other hand, the original source of the oscillations is not one single point (as it was for Figure 3.9) but an extended area or volume of a size that is large compared with the wavelength of the waves propagating from it (Figure 3.10), we have a more complicated superpostion to perform, since each point within this source will then produce a pattern similar to Figure 3.9 and all these patterns will have to be superposed to get the resulting wave pattern. Under these circumstances one can show (again, we will not do this in detail) that the interferences combine so that out-

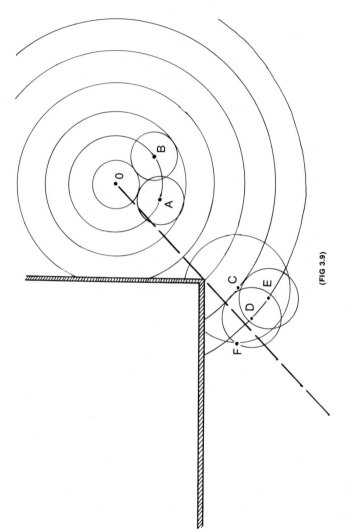

(FIG 3.9)

Figure 3.9 Constructing the diffraction of waves around the corner. The original source of the wave is at 0. From it circular waves propagate, constructed as explained in Figure 3.8. Even though point *F* is "around the corner" from point 0, the interference of the waves from *C*, *D*, *E*, and many others will create a wave there also.

side the line of sight to the source, the waves will produce an almost complete destructive interference, and hence the resulting wave will have an almost zero amplitude around the corner. "Almost" here means that just a very little around the corner the amplitude will still be finite though tiny, but upon going a bit farther around the corner, the amplitude will be, for all practical purposes, zero. In Figure 3.10 only the first few waves are indicated around the source, which is taken to be a segment of a straight line (though this is by no means necessary, we could have taken a source of any other shape). We see that the wavelength is much smaller than the source. In the area in which none of the source is in line of sight, the interferences will be, for all practical purposes, completely destructive.

It is important now to elaborate on the remark that even for a source much larger than the wavelength, there will be a small amount of propagation of the wave even around the corner over a small distance. More precisely, the distance around the corner over which such an extended source will propagate is about one wavelength of the wave. This is important when we are not talking about propagation around the corner but about propagation behind an obstacle of finite size that is in the path of the wave. We just saw that the wave can propagate "behind" (that is into the "shadow" of) such an object, to a distance of about one wavelength. If, then, the size of the obstructing object is much less than one such wavelength, the wave will penetrate the area behind the object from both the right and the left, and in fact the interference of these two sideways penetrations produces a wave almost exactly the same as if there were no obstructing object at all. If, on the other hand, the obstructing object is much larger than one wavelength, the penetration will show up only at the edges of the object, and most of the area behind it will be in a "shadow" from the waves. We can think of the situation, in a somewhat anthropomorphic way, as if the waves had "eyes" of the size of their wavelengths, and these eyes could "see" objects comparable in size to the eyes but would not see objects that were tiny compared with the size of the eye, just as we humans can see things only if they are "sufficiently" large, that is, if they are at least approximately comparable in size to our human dimensions.

Now we can understand the examples for diffraction processes given earlier. Light has a wavelength (as we mentioned) that is tiny compared with the size of everyday objects around us, including the size of the light sources, and so, as we saw in Figure 3.10,

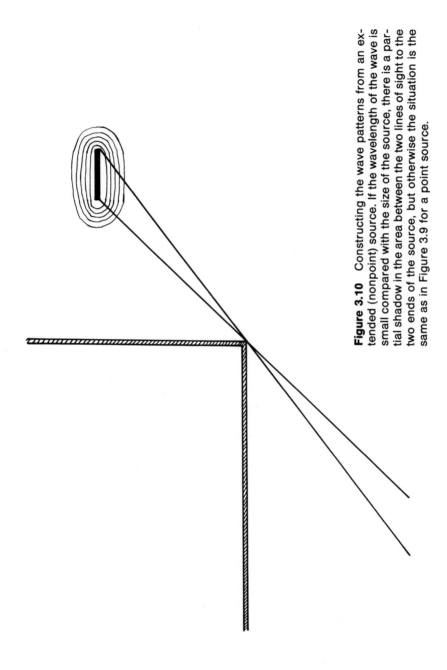

Figure 3.10 Constructing the wave patterns from an extended (nonpoint) source. If the wavelength of the wave is small compared with the size of the source, there is a partial shadow in the area between the two lines of sight to the two ends of the source, but otherwise the situation is the same as in Figure 3.9 for a point source.

light will not "diffract" around the corner but casts a sharp shadow. On the other hand, the AM radio wave, with a wavelength of 1000 feet or so, will easily diffract around relatively small objects such as a tree, a house, or even a small hill. This is not the case for FM radio waves with a wavelength of only 10 feet or so, and so for FM reception we need to be much more in a line of sight with the source of the emission of these waves. The post driven into the ocean floor has a very small diameter compared with the wavelength of ocean waves (eight to 10 inches *versus* 10–20 feet) so that the waves easily diffract around the pole. They do not, however, diffract around a pier or wall that is 50 feet long, and hence the boats moored behind it are protected from the stormy waves even if the harbor is not completely enclosed.

Summary

In this chapter we learned that oscillations can be spread over an extended spatial area or volume if the individual oscillating points are linked together so that as one point oscillates, the neighboring ones will also be forced to oscillate, and then the next neighbors, and so on. If all these oscillations have the same phase, we describe a standing wave, because the crest does not appear to move with time. If, on the other hand, any two neighboring points have the same nonzero phase difference in their oscillations, we talk about traveling waves, because the crest of the wave appears to move with time (even though the individual points that oscillate do not have an overall motion apart from their periodic oscillations).

If the oscillations of a given point are in the same direction as that of travel of the wave, we call it a longitudinal wave; if the two directions are perpendicular to each other, we speak of a transverse wave. In gases only longitudinal waves are possible.

Waves are characterized by frequency and wavelength, the product of two being equal to the speed of propagation of the wave. A wave whose amplitude decreases with time is called damped. Those points along a standing wave that are *always* at rest are called nodes.

When a wave reaches the boundary of two materials, part of it reflects from the boundary and travels back in the medium from which it came, and part of it passes the boundary and travels on in the other medium. If the wave hits the boundary in a direction not perpendicular to the boundary, the reflected wave will bounce back at the same angle at which it came in, but the transmitted wave will travel in the second medium at an angle that is different from the angle of incidence. This phenomenon pertaining to the transmitted wave is called refraction.

Besides sound waves, which will be the main subject of our upcoming discussion, in everyday life we commonly encounter electromagnetic waves, such as light, radio waves, and x-rays. These can also propagate in empty space, whereas sound waves require a material medium for propagation.

Waves can also bend around an obstacle in their path, a phenomenon called diffraction. This can come about because of the interference of waves emitted from various parts of the source and of the air around it, and is noticeable if the size of the source is small compared with the wavelength of the wave. If the source is large compared with this wavelength, the various waves will interfere so that they do not bend out of the line of sight, and hence throw a sharp shadow. The sharpness of the shadow (that is, the transition between shadow and no shadow) is about the width of one wavelength of the wave. Since the wavelength of visible light is tiny compared with the usual everyday dimensions of objects, light almost always throws a sharp shadow, while the wavelength of sound waves is comparable to the size of everyday objects and so sound waves are usually also heard around the corner.

Part **II**
The Sound Getting To You

Chapter Four

Propagation of Sound Waves

FOLLOWING OUR GENERAL study of oscillations and waves, we turn to some specific aspects of sound waves. To do so we have to ascertain the relationship of the *physical* characteristics of waves to the *everyday* characteristics we ascribe to sound, as is done in Table 4.1.

The second and third of these aspects of sound will arise in our discussions in subsequent chapters. In this chapter we will mainly consider loudness, and one aspect of pitch.

Loudness

Experimental studies have shown that the loudness of sound is related to the amplitude of the sound oscillations. The louder the sound is judged to be, the larger the amplitude is found to be. The relationship between the two is *not* a simple proportionality. The question of what we mean, as humans, by "twice as loud" is not even well defined, as we will see in a later chapter. What we can measure, however, in an instrumental and objective way, is the total amount of *sound energy* that reaches a certain spot as a

Table 4.1
Characteristics of Sound Waves

Everyday Description of the Perceived Aspect of Sound	*Corresponding Physical Characteristics*
Loudness, volume of sound	Amplitude, the amount of energy in the oscillation
Pitch (high or low)	Frequency, generally that of the lowest Fourier component
Timbre, the kind of sound (i.e., noise, singing, drum, violin, etc.)	Fourier spectrum of the oscillation and the time evolution of the wave

result of sound waves propagating to that spot. It can be determined that this sound energy is proportional to the *square* of the amplitude of the sound wave. In what follows we argue primarily in terms of this sound energy.

It is our everyday experience that the further we are from a sound source, the softer we hear it. It is this relationship that we now explore in a scientific way. Since we know that our subjective perception of loudness is related (even if not in a simple proportional way) to the amount of energy of the sound wave at a given point, the observation of the relationship between distance and loudness should be explainable in terms of the relationship between distance from the sound source and the energy of the sound wave.

Consider a point source of sound: a small but powerful firecracker, or something similar. The source is located at point 0 (Figure 4.1). We will assume that the sound waves propagate the same way in all directions from 0. We will then measure the amount of sound energy falling on a detector that has an area of unit size (for example, one square foot or one square centimeter). We do such measurements at various distances from the source. How will the amount of energy falling on the detector change with distance?

The answer to this question is made very easy by the law of the conservation of energy. As the sound wave travels away from the source, the *total* amount of energy this wave carries must remain the same. In saying this we assume that as sound travels through air, very little of its energy is dissipated into heat in the air. In reality there is *some* loss into heat, and so our investigation here will be only approximately valid.

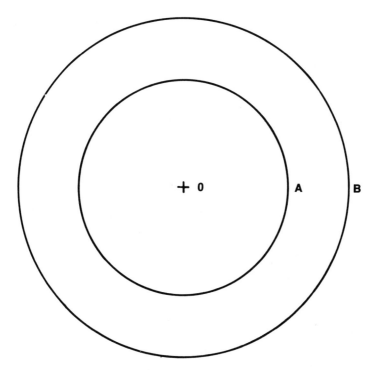

Figure 4.1 The sound wave originating at point 0 at later times is spread over the surfaces of spheres *A* and *B*.

Thus we will use the fact that when the sound is distributed on the surface of the sphere *A* (Figure 4.1), the *total* amount of energy it carries is the same as when the sound wave has reached sphere *B*. The energy *per unit area* (which is what we measure with our detector) is, therefore, on each sphere, the same amount of energy divided by the surface area of *that* particular sphere, and since sphere *B* clearly is larger than sphere *A*, the energy per unit area clearly will be smaller on sphere *B* than on sphere *A*.

Thus, qualitatively, we have explained why sound farther from the source is softer than sound close to it. But we want to do better than that, and make a *quantitative* prediction for how the energy per unit area varies with the distance from the source.

It may appear that to do so we need the formula for the surface of a sphere. Upon hearing this, you rummage in your mind for that formula you had to memorize for your high school teacher but have most likely forgotten. While trying to recall the formula you blame me for the apparent breaking of my initial promise that we will not use any formulas and will not need any "prerequisites" for this course.

Well, do not stretch your memory for the formula. You do *not* need a formula for figuring out the decrease of the energy per unit area. Instead we will work with a much more beautiful, general, and universally useful concept that quickly will provide us with the answer. This concept is called "scaling," and even though you probably have never heard of it, you will understand it shortly and thereafter will be unable to forget it.

Scaling

Let us start with a simple example, a rectangle with its two sides 2 cm (centimeters) 3 cm respectively (Figure 4.2a). That the area of this rectangle is 6 cm^2 can be demonstrated by drawing squares, 1 cm on the side, into the rectangle.

Now let us consider a rectangle that, sloppily speaking, is "three times as big," by which we usually mean the *each side* is three times as long (Figure 4.2b). One can then again draw squares 1 cm on the side into this rectangle, and thus determine that its area is 54 cm^2. This is nine times as much as the area of the small rectangle. We can see that *along each side* there are three times as many squares, and hence over the whole surface there are three times three (or 3^2) times as many squares. Since we are considering surface, which is a *two-dimensional* property of the rectangle, enlarging the rectangle by a given factor along *each side* will produce an enlargement of this two-dimensional property by *the square* of that factor.

For simple, regular shapes such as rectangles this way of figuring the area offers not much of an improvement over conventional methods using formulas, but our new method becomes very powerful if we apply it to irregular shapes (Figure 4.2c, d). If we ask how many times larger the area of the figure in Figure 4.2d is than the area of the figure in Figure 4.2c when *each dimension* of the latter was enlarged by a factor of three to get the former, our reasoning is the same, and the answer remains "nine." Note that this method cannot be used to calculate the actual area of either of the figures. We can use it, however, to predict the *ratio* of the two areas, *independently* of the knowledge of what their areas are. This is what gives this method such great power, since calculating the area of either of the figures (Figure 4.2c, d) would be quite difficult. The *ratio* of the two, however, depends only on the *dimensionality of the concept of area*, that is, only on the fact that area is a *two*-dimensional concept.

Note that this method can be used only if we have two figures that are *identical in shape* but of different sizes. The method in general would break down for figures with different shapes.

Let us practice our newly gained technique on other examples. What if we ask not about the area of these figures, but about the length of their perimeter? For example, we have a plot of land of rectangular shape, 200 feet by 300 feet, and another plot 600 feet by 900 feet. How much more fencing is needed to enclose the second plot than the first plot?

As length is a *one*-dimensional concept, we can confidently answer that the larger plot which in each dimension is three times as large as the smaller one, will need three times as much fencing. The same would be true if the two plots had identical but irregular shapes of the sort shown in Figure 4.2c and d.

For further practice, consider a reservoir, one mile long, one-half mile across, and 50 feet deep, and another reservoir two miles long, one mile across, and 100 feet deep. Apart from the sizes, the two reservoirs have the same shape. How many times as much water can be stored in the larger reservoir than in the smaller one?

As volume is a *three*-dimensional concept, and the larger reservoir is twice as large as the smaller one in each dimension, the larger reservoir can hold 2^3 = eight times as much water as the smaller one.

This method of calculation is called "scaling," because we take a smaller figure and "scale it up" to compare with the larger one, or we take a larger figure and "scale it down" to compare with a smaller one. Scaling has an enormous variety of interesting applications in everyday life, including the following.

1. An elephant has a much thicker leg *relative to the size of its body* than does the mosquito. This is so because the leg must carry the weight of the body, and while the strength of the leg is

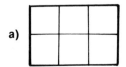

a)

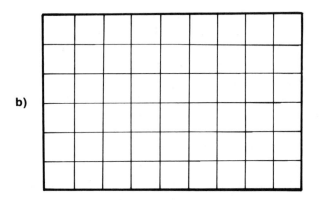

b)

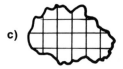

c)

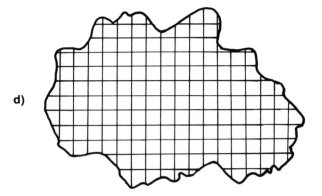

d)

Figure 4.2 Scaling is illustrated by the comparison of the rectangles in (a) and (b). *Each* side of that in (b) is three times longer than the corresponding side for that in (a), but the shapes of the two rectangles are similar. Hence the area (a two-dimensional quantity) of (b) is $3^2 = 9$ times larger than that of (a). Such scaling works also for irregular shapes [see (c) and (d)], for which the calculation of the areas is complicated. The *ratio* of the areas is, however, still 9.

determined by its cross-sectional area (two dimensional), the weight of the body is proportional to its volume (three dimensional). Thus a mosquito, scaled up in each dimension, say, by 1000 to have the same height, length, and width as the elephant, would weigh 1000^3 = one billion times what it does now, but its legs would have a cross-sectional area only 1000^2 = one million times what it is now. Thus, if the legs were just adequate for the real mosquito, they would be inadequate by a factor of 1000 in the scaled-up version to carry the scaled-up weight of the mosquito.

2. Other things being equal, a somewhat larger city requires a considerably larger number of commuting roads leading into it from its suburbs than does a smaller city. This is so because the road is a one-dimensional object, but the number of workers in a city depends on its area, which is a two-dimensional quantity.

3. In today's very large jet planes, it is more difficult to get a window seat than it was in the old, small, propeller planes. This is so because the number of window seats depends on the perimeter of the plane (one dimensional) while the number of people fitting in a plane depends on the area of the compartments (two dimensional).

4. From the point of view of heating, large apartment houses are much more efficient than individual family homes. This is attributable to a large extent to the fact that the number of people living in a house is proportional to the volume (three dimensional) of the house, while the heat loss from the inside of the house to the outside is proportional to the surface area (two dimensional) of the house.

Dependence of Sound Energy on Distance

Armed with our powerful technique of scaling, we now return to our problem of determining the dependence of sound energy per unit area on the distance from the source. We first take our point sound source in an environment in which the sound travels equally unobstructedly in all directions away from the source. We worked out this problem up to the point when we had to determine how many times larger the surface of sphere B is (Figure 4.1) than the surface of sphere A. From scaling we know that since the two spheres are identical in shape (both are spheres), and the radius of sphere B is 1½ times that of sphere A, its area will be $1½^2 = 2.25$ times larger than the area of sphere A. Note that we could determine this ratio without having to recall or reconstruct the formula for the surface area of a sphere.

Since the surface area of sphere B is 2.25 times that of sphere A, the energy per unit area on sphere B will be 2.25 times less than on sphere A, by our previous argument, based on the fact that the *total* amount of energy distributed over sphere B is the same as that over sphere A. In general we can then see that as we increase our distance from the source by a factor, the sound energy per unit area will decrease by that factor squared. Thus at four times farther we will measure only 1/16th the sound energy per unit area.

This holds for the situation in which sound can travel away from the source *in all directions equally*. Consider, in contrast, a very different situation of a long tunnel in which we place a sound source at point 0 (Figure 4.3). We will assume that the walls of the tunnel are perfectly reflecting, that is, they reflect all the sound waves falling on them and absorb or transmit none. This is only an approximation of a real-life situation, but a good one if the tunnel walls are hard rock. In this case how will the sound energy per unit area change as we go from place A to place B in the tunnel?

We see that the area over which the total sound energy is distributed in this case is the cross-sectional area of the tunnel, which is the same at A as it is at B. Hence by the same argument we used earlier, the energy per unit area will be the same at B as it is at A, and we will hear the sound source equally loud whether we are closer to or farther from it.

As a third situation, consider a huge roofed shed, open on the sides, at the center of which a band is playing on a platform, and people are sitting all around the band (Figure 4.4). How will the sound energy per unit area reaching the listeners vary for the listener sitting closer to and the listener sitting farther from the band?

We will assume that both the floor and the roof of the shed are perfectly reflecting. In that case the sound energy is equally distributed over the area of the perimeter of a cylinder with a center at the band, and a fixed height, which is the height of the roof. As we go farther and farther, the area of this cylinder grows only with one of the two dimensions (the perimeter of the cylinder), because the

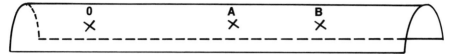

Figure 4.3 Scaling in a tunnel. Since the tunnel is a one-dimensional object (if its length is much greater than either its width or height), the loudness of sound at *B* will be the same as at *A*, since the cross-sectional area of the tunnel over which the sound is distributed is the same at *A* as at *B*. (We neglect the possible absorption of sound on the walls of the tunnel.)

other dimension, which is the height of the roof, remains the same, and so we are dealing here with a one-dimensional quantity. Hence the sound energy per unit area will decrease by the same factor as that by which the distance from the band increased.

As a last example, consider a flat countryside, out in the open, with a sound source at ground level. How will the sound energy per unit area vary as we go farther from the source? We see from Figure 4.5 that here also we have two identically shaped bodies (hemispheres), and hence the surface areas will grow as the linear dimensions (the distance between the source and the observer) squared. Thus going three times as far will produce one ninth of the sound energy per unit area. The situation is the same as if we had a sound source from which sound could travel in *all* directions, and not only in half of the directions, as it is the case for our example.

The results derived here hold, strictly speaking, only in the idealized situations of the sound propagating in all directions equally, reflecting from surfaces totally, not being absorbed in air at all, and so on. Thus the results give only an approximate idea of the sound intensity as it travels. We discuss some deviations from these idealized results in Chapters 13 and 14.

The Speed of Sound

For a further discussion of phenomena related to the loudness of sound, we will now review a few items concerning the speed of

Figure 4.4 Scaling in a covered shed. If the band plays in a cylindrical covered shed and the absorption of sound by the floor or the roof is small, the sound intensity will be inversely proportional to the *first* power of the distance from the band, since the sound is distributed over a surface (a cylinder drawn around the band) of which only *one* dimension (the diameter) changes as we go farther away from the band.

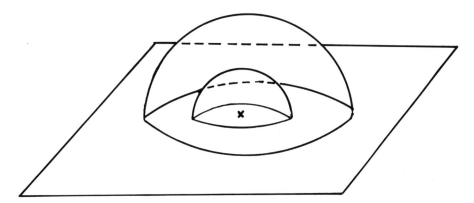

Figure 4.5 Scaling in the open air, with the source located on the ground. Since the sound is distributed over a hemisphere that is just as two dimensional as a whole sphere, the drop-off with distance will be the same inverse square as for a source suspended in space.

sound. As we mentioned earlier, the speed of sound in air is about 1000 ft/sec, or about 600 mi/h.

The first observation concerning this value is that it is relatively small, compared to the kind of speeds we can produce artificially or observe in everyday life. This is unlike the case of the speed of light, which is 186,000 miles per second, and so is enormous compared with virtually anything we experience in everyday life. Note, however, that in the hi-tech space age, there are now exceptions. Your intercontinental telephone call goes via a synchronous satellite located about 22,000 miles from the earth, and as the telephone signal is electromagnetic, it travels with the speed of light. Thus when you have finished your sentence, and await a reply, a distance of four times 22,000 miles must be covered by the message and the reply together, which adds almost half a second delay to the normal delay (which is caused by a person's physiological reaction time). This extra half a second is noticeable, in that you might get the impression that your European friend is not quite as "sharp" as when you met in person in Paris last summer. Incidentally, if such a telephone conversation were not transmitted electromagnetically, but through direct sound waves (periodically amplified so that the loudness would be adequate), the reply would come to New York from Paris, not a half second late, but 14 hours late. This also gives you a feeling for how much faster light travels than sound.

But with regard to sound, since we frequently encounter in everyday life speeds that are comparable to, or even in excess of, the speed of sound, there are some conspicuous phenomena that are related to the attainment of speeds equal to or in excess of the speed of sound. We will look at two of those.

The Sonic Boom

One phenomenon is commonly known as the "sonic boom." Consider the sound emanating from a sound source that is at rest relative to the air around it (Figure 4.6a). This is a snapshot of the crests of waves around the source, taken at a given time, and the circles (actually spheres) around the point show the crests of the waves that were emitted from the source at various times *before* the time of the snapshot. Naturally, the longer ago the particular crest was produced in the source, the farther it will have spread away from the source. The numbers labeling the spheres give the number of seconds *before* the time of the snapshot when that particular crest was produced in the source.

Figures 4.6b–d show similar snapshots, but for a source that is constantly moving with respect to the air. You see that in those cases the various circles no-longer are concentric, since the position of the source is different at different times, as also shown. Figures 4.6b–d show this for various speeds of the source compared with the air. Figure 4.6b is for a speed of the source, which is about one-third of the speed of sound. Figure 4.6c shows the case where the source moves exactly with the speed of sound, and Figure 4.6d the case when the source moves about 1½ times as fast as the sound does.

Note that in Figure 4.6c all the previous wave crests overlap, at the time of the snapshot, with the location of the source at that time. In other words, since the sound source travels at the same speed as the sound itself, the source always keeps even with the sound wave it emitted in the forward direction at a prior time, *whatever this prior time may have been.* This piling up of the sound waves and the energy contained in them contributes to the technological difficulty we once had in "breaking the sound barrier."

The situation is similar but different in detail in Figure 4.6d where the speed of the source is even greater than the speed of sound. In that case the overlap of crests at the time of the snapshot occurs in the area marked by the two parallel straight lines, so there is, at a given time, a whole "wavefront" where the crests overlap. Because of this overlap, the sound there is very loud; it causes a sonic boom when the source is a supersonic plane. Of course, as the source moves on, the area of the wavefront moves with it, so that a snapshot a little later would show the same figure moved a bit to the right. Therefore, at any given fixed place on the ground, the sonic boom is a very loud but very short bang.

Speed of Sound in Different Media

The speed of sound in all media is *roughly* the same order of magnitude as in air (meaning that it is within 10 times larger and 10 times smaller). In looking at it more closely, however, we see considerable differences within this common order of magnitude. Table 4.2 gives the approximate values for some common materials. We see that the speed tends to be greater in solids than in gases.

Differences also exist with regard to whether the wave is transverse or longitudinal. In air and the other gases, there is no choice, as we saw, since all waves in gases are longitudinal, but in liquids, and particularly solids, we can have either.

The speed of sound in a *given* medium also depends on the density of the material, since the ease of transmitting the oscillation from one molecule to another can be influenced by how close these molecules are to each other. The speed also depends on the elasticity properties since, again, the propagation of oscillations from one part of the material to another is influenced by the elastic properties. In gases the effect of "elasticity" (corresponding there to pressure) approximately cancels the effect of density, and so the speed of sound in gases does not depend very much on density. It does depend, however, on temperature (which, as we saw, was the measure of the individual kinetic energies of gas molecules). The

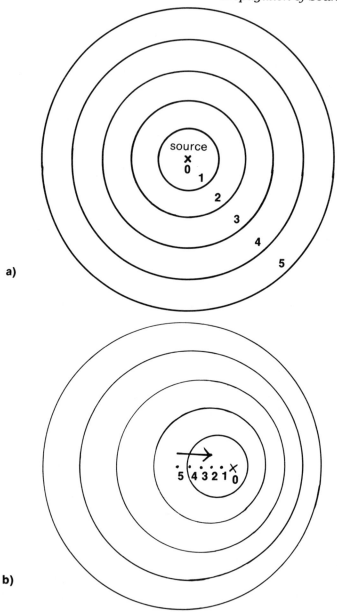

a)

b)

Figure 4.6 Illustration of how shock waves (sonic booms) come about. Figure 4.6a shows the wave locations at various times (0 second, 1 second, etc.) after the sound is produced at a stationary point marked "source." The wave locations form concentric circles. Figure 4.6a is a snapshot of the various waves produced at different times: the wave on circle 5 produced five seconds ago, that on circle 4, four seconds ago, etc. In Figure 4.6b the source is no longer stationary but moves with time. The dots indicate the position of the source five, four, etc., seconds ago, and the position of the wave emitted, say, three seconds ago is the circle marked "3." We see that the wave locations are circles but no longer concentric. They bunch up in the direction in which the source moves.

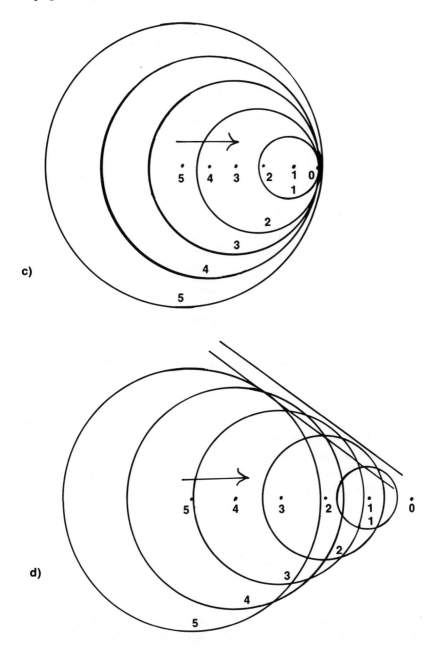

Figure 4.6 (continued) In Figure 4.6c the speed of the source is the same as the speed of propagation of the sound, and hence the "bunching up" toward the front is now extreme. In the foreward direction, all sound arrives at the same time, no matter when it was emitted and from where. In Figure 4.6d the speed of the source is larger than the speed of propagation of the sound, and consequently the bunching up now occurs along two strips at an angle to the direction of motion of the source.

Table 4.2
Speed of Sound in Various Media Under
Various Circumstances (units of meter/second)*

Air	330	Steel	5000
Oxygen	320	Iron	4500
Nitrogen	350	Nickel	5000
Water vapor	400	Copper	3600
Carbon monoxide	340	Brass	3500
Carbon dioxide	260	Aluminum	5100
Helium	930	Lead	1200
Hydrogen	1300	Tin	2700
		Gold	2000
		Silver	2600
Water	1440	Tungsten	4300
Alcohol	1200		
Benzene	1200	Diamond	14000
Turpentine	1400	Ivory	3000
Mercury	1400	Glass	5000
		Brick	3600
		Cork	500
		Rubber	50
		Maple ⎫	4300
		Pine ⎬ Along the fiber	3300
		Oak ⎭	3800

*The values are approximate. The speed depends on temperature. The values given here are for about 1–20°C.

higher the temperature, the faster sound travels, since the faster speed of the individual molecules decreases the time between collisions of molecules, and these collisions are responsible for propagating the longitudinal wave.

The Doppler Effect

When a car approaches you and the driver blows his horn, you hear a certain pitch of sound. As the car passes by, the pitch suddenly drops, and then stays at this lower level. Since it is unlikely that the driver adjusted the horn suddenly and for your benefit, the phenomenon must be an "apparent" one, connected with the motion of the car. Indeed the driver, listening to the pitch of the horn on the car, would at all times hear a pitch that is halfway between the two pitches you hear, before and after the car passes you.

Pitch, as we mentioned earlier, is connected with the physical property of frequency, that is, the frequency of the wave reaching your ear will be referred to by you as the pitch of the sound. "High" pitch corresponds to larger frequencies, and "low" pitch to smaller frequencies. The terms "high" and "low" pitch that we use in everyday language have nothing to do with the other meaning of these words, to denote large or small altitudes. You could, with equal justification, call large-frequency sound waves red pitches and the small-frequency sound waves blue pitches. In semitechnical language the two terminologies are even more confusingly merged, and we talk about "high frequencies" (meaning large frequencies) and "low frequencies" (meaning small frequencies). In deference to this common practice, we also have to use these ambiguous terms.

To understand how these apparent rises and drops in pitch occur, we proceed in two independent ways: First we use an analogous physical situation, which argues in terms of frequency, and then we argue in terms of wavelengths, using some of the figures we have already established for a different purpose.

An Analogy

A young man is located at *A*, and a woman friend at *B*. The distance between *A* and *B* is 240 miles. To communicate, the woman releases a letter-carrying pigeon every day at noon. The pigeon flies at the speed of 10 mi/h straight in the direction of *A*.

Let us first assume that both the man and the woman stay at their respective positions (Figure 4.7a). How often will the man get news from the woman? The pigeon released on day 1 at noon at *B* will arrive at *A* on the next day at noon, since at the speed of 10 mi/h it will take the pigeon 24 hours to cover 240 miles. Thus the pigeon of day 1 will arrive at noon on day 2, the pigeon of day 2 will arrive at noon on day 3, and so on. The man will get news every 24 hours, that is, at the frequency of exactly once a day, which is the same as the frequency with which the pigeons are released at B.

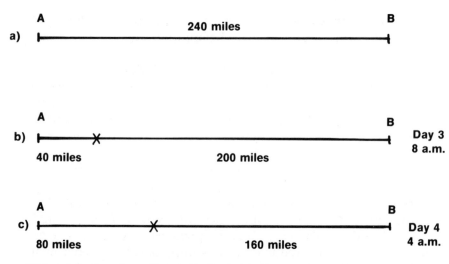

Figure 4.7 Doppler effect as illustrated in the text by communication between two people via letter-carrying pigeons. The sender of pigeons at *B* is stationary. If the receiver of pigeons also remains stationary at *A,* and the pigeons fly at 10 mi/h, *A* will receive pigeons sent from *B* every day at noon on the next day at noon, 24 hours apart. If the receiver walks toward *B* at 2 mi/h, starting at noon just after receiving a pigeon, he will receive the next pigeon earlier than 24 hours later, and at a different position [in dicated *x* in (b)]. The day after, he will receive the next pigeon again earlier and again at a different spot, indicated by *x* on (c). Thus the walking receiver will receive pigeons not at a frequency of one per day, but more often than that (as it turns out, once every 20 hours).

Now let us assume that the man becomes impatient and starts the long walk toward *B*. Let us also assume that he walks at a steady speed of 2 mi/h. He starts walking on day 2 at noon, just after he receives the pigeon sent off from *B* at noon on day 1. When will he receive the next pigeon? (See Figures 4.7b and c.)

It is not difficult to realize that he will receive the next pigeon on day 3, but *before* noon, since he is walking in the direction from which the pigeon will come, and hence the pigeon has to fly a shorter distance to reach him. In particular, let us consider day 3, 8 a.m. The man will have walked for 20 hours, at 2 mi/h, and so has reached a point 40 miles from *A* toward *B*. The pigeon, which was released from *B* on day 2 at noon (in other words, at the same moment the man started his walk from *A*), will also have flown for 20 hours by 8 a.m. on day 3, at 10 mi/h, so the pigeon has covered 20 times 10 = 200 miles and will now be 200 miles from *B* toward *A*. We see that at this time (8 a.m. on day 3) the man and the pigeon meet. *Only 20 hours* have elapsed since the man got his last news, as compared with the *24 hours* when he stayed at *A*. Thus when he walks, he receives news via the pigeons every 20 hours, that is, at a larger frequency than before. (Previously the frequency was exactly once a day; now it is 1/(20 hours) = 1/(5/6 of a day) = 6/5 per day.)

Note that the ratio of the speed of the man to the speed of the pigeon is (2 mi/h)/(10 mi/h) = 1/5. Note also that the new frequency is 6/5, or 1 + (1/5). This is a general rule:

New frequency is the old frequency plus the old frequency multiplied by the ratio of the speed of the recipient of the wave to the speed of the wave itself. Or we can say that the *frequency difference* between the new and old frequencies is given by the old frequency multiplied by the above ratio of the two speeds.

In our example the pigeon was the "wave" (since it carried the signal), and the man was the *moving recipient* of the wave.

Had the man decided to walk *away* from the woman, the pigeon would have required *additional* time to catch up with him, and hence the time elapsed between two messages received by the man would have been *more* than 24 hours. In that case the new frequency would be the old frequency *minus* (instead of plus) the old frequency multiplied by the speed ratio, or the frequency difference would have been negative (that is, the new frequency would have been smaller than the old). We see from this that the "normal" frequency, one that is heard by an observer who is stationary with respect to the source of sound, is halfway between the two pitches being heard by an approaching recipient and a receding recipient.

In this example it is the recipient who is in motion and not the source, as it was in the example of the moving car and its horn. Thus the pigeon analogy would instead correspond to the apparent drop in pitch of the bell at a railroad crossing as heard by somebody on a passing train. But it does not matter which of the two (recipient or source) is at rest and which is in motion. Only the *relative* speed between the two matters.

Another Way to Understand the Doppler Effect

If we can show, instead of arguing directly on the basis of *frequencies*, that the *wavelength* of the wave reaching the stationary recipient from a moving source changes as a result of the motion of the source, we can also conclude something about frequencies, since, after all, wavelength and frequency are related in that their product is equal to the speed of the wave in that medium.

But we can certainly argue in terms of wavelength if we again inspect Figures 4.6a and b. On the former we can literally see the wavelength from the snapshot: It is the distance between the circles, since each circle shows a crest of the wave at that particular moment. If we then compare this wavelength with the wavelengths we can read off Figure 4.6b, we see that in the *forward* direction (i.e., when the source and the recipient approach each other), the wavelength of Figure 4.6b is smaller than that of Figure 4.6a, while in the *backward* direction (when the source and the recipient recede from each other), the wavelength increases compared with Figure 4.6a. A *de*creasing wavelength must result in an *in*creasing frequency (i.e., in a rise in pitch), while an *in*creasing wavelength will go with a *de*creasing frequency, a drop in pitch. Thus we arrive at the same result for the Doppler effect as we did in the pigeon analogy. It can be shown (see exercise problem) that this second approach also gives the same result *quantitatively*.

Summary

After connecting loudness, pitch, and timbre with various physical characteristics of waves, we explored the relationship between distance from the sound source and loudness, and found that the amount of sound energy per unit area decreases as the square of the increasing distance from the source. We arrived at this conclusion by using the concept of scaling, a very simple and broadly applicable idea in science. The relationship holds for a sound source from which sound can emanate in all directions equally. Other types of situations were also analyzed, using scaling.

If a sound source moves in air at a speed that is somewhat greater than the speed of sound itself in air, an accumulation of wave crests occurs along a wavefront, causing sonic boom. The speed of sound is different in different media, although its value is within a factor of 10 or so on either side of the average. This speed is very much less than the speed of light.

A phenomenon in which the ratio of the speed of a moving sound source (or of a moving recipient of sound) to the speed of sound itself is important is the Doppler effect, which we understood in two ways: through an analogy of a letter-carrying pigeon and a moving recipient of the letters, and through the study of spherical waves emitted from a moving source. The shift in frequency is equal to the ratio of the two speeds multiplied by the original frequency; it is positive (the pitch rises) when the source and the recipient approach each other, and negative (the pitch falls) when the two recede from each other.

Chapter Five
Perception of Sound

AT THIS POINT in our study of sound it is appropriate to say something about how we humans perceive and register sound waves approaching us. There are many applications of sound-type waves, of course, where the measurement and registration of sound are not done through the innate organs we have for that purpose. For example, high-frequency sound waves are used for therapeutic purposes in medicine, where instruments measure the correct dose, as well as the effect of the waves. When measuring seismic waves, we also use instruments. Even in the manufacture of musical instruments, instrumental analysis is used in conjunction with expert aural opinion. The same is true in building auditoriums or in testing "hi-fi" equipment. Yet for many purposes for which you are interested in learning about sound and music, the receptor and arbiter will be you, that is, your ears and brain.

Note that we say "ears *and* brain." This is the first important point to make about the perception of sound. Even if we have completely understood the physics and the physiology of the ear, we have mastered only part (and probably the smaller part) of the problem of perception. What is transmitted from the ear to the brain can then be interpreted by the latter in a multitude of complex, sophisticated, ambiguous, and subjective ways, a process that is, at least at this stage of our understanding, mostly outside the realm of physics, and in fact barely within the realm of science, if at all.

It is not difficult to demonstrate this just by referring to some familiar situations. Some claim to "hear voices," even in the complete absence of a physically measurable auditory stimulus. Virtually all of us have been spoken to in our dreams, or have heard noises during such dreams. If you spell a name or give a telephone number to another person, that person will often interchange two letters in the spelling or two digits in the phone number, claiming that he "heard it that way." When conversing at a noisy party, you can pick out the speech of the person to whom you are speaking over an equally loud background noise, if you concentrate.

Since the story of human sound perception is so tied in with nonphysical factors that we do not yet understand, our discussion of this subject will not be as neat, complete, logical, and easily retainable as some of the topics covered here. Nevertheless a considerable portion of this subject matter can be discussed systematically on the basis of the concepts of physics we have learned, especially in connection with the structure of the ear itself. That is the main objective of this chapter.

How the Ear Works

In most books on the physics of sound, the structure of the ear is discussed in terms of anatomically accurate diagrams of the cross section of your ear and skull, replete with anatomical names in Latin thus making it appear to be a complicated subject to be memorized and not understood. To be sure, if you are going to be an ear surgeon, you will need these details, and you will undoubtedly have to learn them in your courses on medicine. For our purposes, however, only the main principles are needed, and they can be understood easily by a simplified model of the ear, in terms of a few important and functional components. The model will reflect the properties of the ear that you would guess must be in evidence. The ear must collect sound waves, and since they are quite faint, should amplify them. It must also analyze the frequency spectrum of the sound. Then the ear must transform this information into nerve pulses that can be transmitted to the brain for perception.

And all these mechanisms must be protected from impact from the outside. This simplified model is shown in Figure 5.1.

We will divide the hearing system into *six components*, as shown in the figure.

The *first component* is the "sound collector." It consists of the externally visible ear, represented schematically by the cone of the figure, and the ear canal leading to the eardrum, where the canal is represented schematically by a tube. The purpose of the collector is to gather sound waves (by reflection) from an area larger than that of the eardrum itself, and to conduct that sound (through the tunnel we discussed in Chapter 4) into your skull where, well protected from external dangers, your eardrum is located.

The *second component* is the eardrum, represented schematically by an elastic membrane. Its purpose is to convert the sound waves in the air to vibrations of something material in the body, namely, the eardrum. Since the edges of the membrane are affixed to harder material in the body, the sound waves impinging on the eardrum are converted into standing waves on this membrane. As you will see, there are three important membranes in the hearing system, all of which have to vibrate sensitively and in complex ways for good hearing.

The *third component* of the hearing system is a transmitter and amplifier of the vibrations from the first membrane (the eardrum) to the second membrane, which we call the linked window membrane. The transmitter consists of a chain of linked bones acting as tiny levers. They are arranged in such a way that they are activated by vibrations over the whole area of the first membrane (eardrum) and cause vibrations on the second membrane (the linked window membrane) over a smaller area but with a larger amplitude. It is in this sense (the conversion from smaller to larger amplitude) that the transmitter is also an amplifier. The *total* sound energy is of course not increased, but the sound energy *per unit area* is.

The *fourth component* is the second membrane (the linked window membrane), which is a window on the fifth component, an enclosed cavity filled with fluid. As this second membrane vibrates, it transmits these vibrations to the fluid in the cavity, and thus creates waves propagating in the fluid (or standing waves generated in this fluid).

The *fifth component* is the cavity (officially called cochlea), with quite hard outside walls except in two places where there are win-

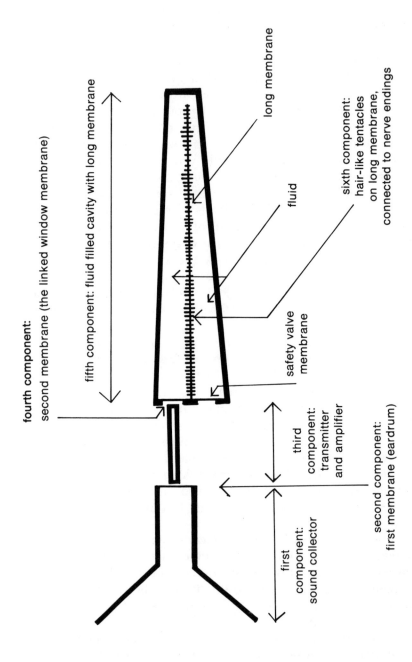

Figure 5.1 Schematic diagram of the human ear, with the six major components discussed in detail in the text.

dows covered by elastic membranes. One of these windows is the fourth component, which, as we saw, is linked with the sound vibrations falling on the ear. The other window, which we might call the safety valve membrane, is there so that when the vibrations are transmitted by the linked window to the incompressible fluid in this cavity, such vibrations can take place, with the safety valve membrane allowing a volume increase of the cavity (by bulging out) when there is a volume decrease by the other membrane bulging momentarily in, and vice versa. The point is that because of the incompressibility of the fluid in the cavity and the hard walls of the cavity, the overall volume must always be the same, and hence any vibration of a membrane window must be "balanced" by a similar vibration of a second window, 180 degrees out of phase with the vibration of the first.

Another important part of the fifth component, (the cavity) is the third membrane in our discussion which stretches the center-plane of the cavity, dividing it into two halves that are completely separated by this membrane, except for a small hole at the extreme right end of the cavity in the diagram. This long membrane has the function of distributing the patterns of vibrations (which are concentrated over the small area of the linked window membrane) over a much larger area, and thus increases the sensitivity to the various details of these vibrations. When the linked window vibrates, it transmits these vibrations to the fluid in the form of waves, and these waves, hitting the long central membrane, set up vibrations all along it.

So far we have achieved the conversion of sound waves in the air into vibrational patterns along a long membrane inside a secluded, well-protected cavity deep inside the skull. The *sixth component* then serves to provide the link between these vibrations and the brain. This is achieved by means of a huge number of very tiny hairlike tentacles located along the long central membrane, on both sides. As the membrane vibrates, the tentacles are made to move by the friction with the fluid. Each tentacle is connected to a nerve ending, which fires a message to the brain about the motion of that tentacle. The brain then integrates the information received from these thousands of nerve endings, to arrive at an overall auditory pattern.

In the form shown here, the mechanism of the ear is easily understandable, is simple, and is uncluttered by details without losing any essential principles. Once this is understood, it will be easy to read other, more technical material, because you will have a "map," an overview of the nature and purpose of the main components.

Spatial Localization of Sound

We have two eyes, and that is what makes it possible for us to localize objects in space by the visual impression we receive. This can be seen by trying to thread a needle while keeping one eye shut. Similarly, the fact that we have two ears located apart from each other allows us to locate the source of sound that reaches us.

There are three different effects that collaborate in giving us this ability to locate aurally (See Figure 5.2). All three depend on S being at a different distance from R and from L.

1. Because of the different distance, the sound intensity (loudness) will be less at L than at R. This is so for two reasons:

    a. The loudness from a given source decreases with the square of the increasing distance, as we saw in Chapter 4.

    b. Ear L is in the shadow of the head, and hence sound gets to it only by diffraction, which may further decrease the loudness as compared with the loudness at R.

2. Because of the different distance, the wave reaching L will have a phase difference compared with the wave reaching R.

3. Because of the different distance, the waves, emitted simultaneously from S, arrive at L and R at different times.

Let us estimate the magnitude of these effects. In the first of the three, let us assume that the sound source is 100 feet away, and that the difference in distances is about one foot. The *ratio* of the two sound intensities (in terms of energy per unit area) would then be $(100/101)^2$, or about 0.98. In other words, we get a 2 percent reduction in loudness between the two ears. That is a small difference (as we will see), which, however, is increased by item 1a. If the distance of the source is only 10 feet, the effect is correspondingly larger, about 21 percent. So we see that it is easier to localize close sources than far ones.

S

source

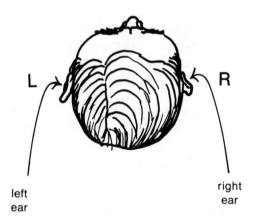

L →   ← R

left
ear

right
ear

S'

source

Figure 5.2 Diagram of how our ability to localize sound in space works. The source S (or source S') is located differently with respect to the right and left ears and hence the two ears receive the sound emitted by the source differently in three respects: the loudness will be different, the arrival time will be different, and there will also be a phase difference between the two waves.

Concerning item 1b, let us remember that diffraction will occur easily if the wavelength is much larger than the size of the obstacle, which in this case is the width of the head, or about one foot. Thus sound waves with large wavelengths (low frequencies) will diffract easier, and so will make spatial location more difficult than sound waves with small wavelengths (high frequencies). Since the speed of sound is 1000 ft/s, the dividing line between the two cases will be, very approximately, at 1000 ft/s divided by one foot, or a frequency of 1000 per second.

With regard to the second effect, the phase difference between the waves reaching the two ears will, of course, also depend on the wavelength. For the frequency of 1000 per second that we just calculated, the wavelength is about one foot, and hence the phase difference between the two ears may be a large fraction of the entire oscillation. For a low-frequency sound, say 100 per second, the wavelength is 10 feet, and so the one-foot difference in distance produces a relatively small phase difference of one-tenth of the entire oscillation, or about 36 degrees. We see that this method of localization also works better for high-frequency sound than for low-frequency sound.

Finally, turning to item 3, the time differential, with the speed of sound 1000 ft/s and the path differential one foot, will be about 1/1000 second. This is quite small compared with the usual times we deal with in everyday life, but experiments have shown that the brain can register and process such tiny time differences and can automatically convert them into a sensation of localization.

It might be remarked that apparently none of these effects differentiate between a source located at S, *in front* of the person, and s', *behind* the person but in a similar relative position. To a large extent this is true, and hence it is much more difficult to distinguish between the locations ahead and behind than between the locations to the right and left. It is, however, not impossible, since the shape of the head and the structure of the ears are not completely symmetric between ahead and behind.

Quantitative Measures of Loudness

We have measured the loudness of sound in terms of the amount of energy per unit area per second contained in the sound wave at a given location. That is an instrumentally well-defined and unambiguous measure. From the point of view of the human perception of the sound, however, simply giving the number describing this energy per unit area is not the best choice, for the following reason.

Experiments on human subjects have shown that the distinctions humans make between sounds is not on the basis of the *difference* in the amount of energy per unit area, but on the basis of the *ratio* in the amount. For example, if you ask a test subject to compare two sounds and tell whether one is louder than the other, the subject will be able to recognize a louder sound if it delivers a certain *percentage* (and not a certain *amount*) more energy than the comparison sound. In other words, if you make the comparison sound quite loud, you have to make the second sound to deliver more energy by a larger amount than if the comparison sound had been rather soft. To put it in numbers, if the threshold of distinction for a human ear between two sound intensities is a ratio of 1.25 (which is roughly the case), then if the comparison sound delivers 50 units of sound energy per unit area per second at a location, the sound that will be judged (just barely) louder will have to deliver 62.5 units. If, however, the first sound had delivered only 25 units, the second, to be judged just barely louder, would have to deliver 31.25 units. The ratio in both cases is 1.25, but the difference in the first case is 12.5 units and in the second 6.25 units.

This observation (that it is the ratios and not the differences that count) is not precise. The experiment is relatively easy if we compare two loudnesses, one being just barely (i.e., on the threshold of being) louder than the other. If, however, we ask a subject to tell us when a sound is "twice as loud as another," we obtain a subjective opinion that can fluctuate from subject to subject, since one's sensation of loudness is not inherently quantified in one's mind, even though comparisons ("louder" and "softer") can be made easily. Nevertheless it is true that ratios approximate the way we perceive loudness much better than differences do. For this reason, the commonly used scale for measuring loudness of sound is based on such ratios. I will now explain how.

To begin let us determine the *range* of energies per unit area per second of sound waves that our ear can register. We can take, as the two limits, something that is just barely audible, and something that is so loud that it is painful to the ear, or even ruptures the eardrum. Measuring those two levels shows that their ratio is extraordinarily large: 10^{14} or so. Since we said that the energy increases as the amplitude of the wave squared, and since the eardrum, which is roughly a half inch in diameter, cannot really vibrate with an amplitude that is much larger than its diameter, we must conclude that when we register a barely audible sound, our eardrum vibrates with an amplitude that is smaller than the maximum amplitude by the square root of 10^{14} or 10^7, that is, with an amplitude that is one ten-millionth inch or less—truly atomic dimensions! The ear is indeed an astoundingly sensitive and versatile instrument.

Our task is to construct a scale that can denote loudness in this enormous range, which is adapted to the fact that it is the ratios that count, and which can measure up to 10^{14} and down to about 1.25 in ratios without having to deal with very large or very small numbers with many zeros. The scale that satisfies all these requirements uses the unit of a decibel (dB). By definition 0 dB is set at the just barely audible level, which then is fixed in terms of a precisely defined number of energy units per unit area per second. From then on you have to *add* 10 dB for each *10-fold* increase in the amount of energy per unit area per second. Note that we *add* decibels as we *multiply* the energy, thus satisfying the spirit of the ratio rule. An easy way to remember the relationship between the number of decibels and the amount of energy per unit area per second is that the number of decibels of a given loudness gives 10 times the *exponent* to which 10 has to be raised in order to give the ratio of the energy per unit area per second at the level of loudness to the energy per unit per second of a barely audible sound.

In technical mathematical terms, such a scale is called a logarithmic scale. You need not, however, understand logarithms in order to understand this description of the scale, or to use the scale. To calculate practical situations to a good approximation, you need only remember the following three rules.

1. Multiplication of energy amounts corresponds to addition in decibels.
2. A 10-fold increase in the energy corresponds to an addition of 10 dB.
3. A two-fold increase in the energy corresponds to the addition of 3 dB.

With these three rules, we can calculate almost anything. Here are two examples.

1. One trumpet in a given spot plays at a certain level and produces 46 dB at some other location. Seven other trumpets now join in at the same spot, all playing at the same level. What will be the decibel rating of the whole ensemble at the same previously indicated location?

   Doubling the number of trumpets always gives three additional decibels. Going from one to eight trumpets means three consecutive doublings (one to two, two to four, and four to eight), so the corresponding operation on the number of decibels is three consecutive additions of 3 dB each. We thus obtain $46 + 3 + 3 + 3 = 55$ dB.

2. What if only four additional trumpets join? The five trumpets can be considered in two steps: first multiplying the number of trumpets by a factor of 10 (making it from one to 10), which corresponds to the addition of 10 dB, thus obtaining 56 dB. Now we will cut down the 10 trumpets by a factor of two to five trumpets. Such a halving corresponds to a *subtraction* of 3 dB, and we end up with $56 - 3 = 53$ dB.

More examples arise in the exercise problems.

Tables of various decibel levels can be found in other textbooks and sources. Just to orient you, a quiet library room has a background noise of 30–40 dB. The usual conversational level is 60–65 dB. Rock music at a disco is often 110–120 dB, close to painful, and in the long run detrimental to your ears.

Note that the decibel scale still measures the same quantity with which we started out, namely, the energy per unit area second. If you have the amount of this energy, you can unambiguously convert it into decibels. The only difference between the two measures is the way in which the two scales are arranged.

This is worth mentioning because various other scales are also used to gauge loudness, which are more oriented toward the details of human sound perception. The "phon" and the "sone" are two of these. The point is that the loudness perception of the human ear also depends on the frequency of the sound wave, and not only on the amount of energy delivered to the ear per unit time. This is clear if we consider some extreme cases: It turns out that sound waves below, say, a frequency of 10 per second are com-

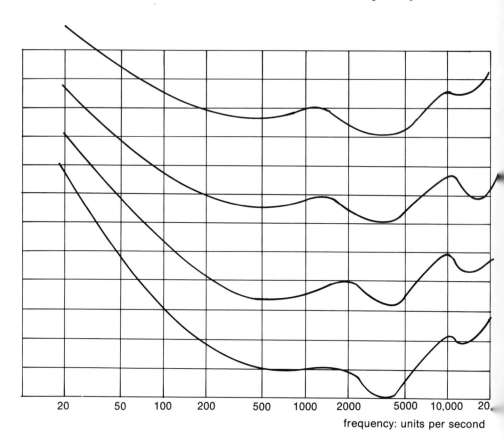

20   50   100   200   500 1000 2000   5000 10,000 20,

frequency: units per second

Figure 5.3   Curves showing equal perceived loudness for various amounts of sound energy (shown on the left-hand vertical scale) at various frequencies (shown on the bottom horizontal scale). For example, a 30-dB sound wave that has a frequency of 100 per second would appear to the human ear as loud as a 10-dB sound wave with a frequency of 1000 per second. We see from the illustration that our ear is most sensitive to a frequency of about 4000 per second, and that as we approach frequencies as low as 20 per second, the sound energy has to be much greater for our ear to perceive it as loud as at 4000 per second. The differences with frequency are greater for relatively quiet sound than for loud sound.

pletely inaudible to us no matter how strong they are, and the same is true for frequencies above, say, 30,000 per second. The zero of the decibel scale is set at the audibility at the most favorable frequency, around 4000 per second. In terms of decibels thus defined, the threshold of audibility is around 20 dB at a frequency of 200 per second, and the same is true at a frequency of 10,000 per second. At even more extreme frequencies, such as 20 per second, the threshold is as high as 70 dB. Figure 5.3 is a schematic indication of how our loudness perception depends on the frequency of the sound. Each line on the figure connects values of decibels (at various frequencies) that would be judged "equally loud" by human subjects.

Such curves also depend on which individual subject we test, and also, in general, on the age of the subject. Infants, for a few months or a year after birth, have a hearing that is much more sensitive to high frequencies than are the ears of most adults, and so high-frequency noise (screeching, whistles, etc.) can alarm a baby although we think nothing of the sound, or perhaps cannot even hear it.

Although frequencies below 10 per second or over 30,000 per second are not "heard" in the usual sense no matter how loud they are, such sounds can still have a perceptory effect. The very-low-frequency sound, if sufficiently loud, will be noticed even though it will not fuse into a steady sound. At the other end, very-high-frequency sound, when sufficiently loud, can do considerable damage and cause pain to our ear and auditory nerves, even though the effect is not likely to be described as auditory from a psychological point of view.

Beats

A special case of an alternation of constructive and destructive interference (discussed in Chapter 2) is the phenomenon of beat, something that is usually judged quite unpleasant to the human

ear. Consider two vibrations with a difference between the two frequencies that is very small compared with the frequency of either
of the two vibrations. For example, Figure 5.4 shows a wave with a
frequency of 12 per second and another wave with a frequency of
13 per second. If we let these two waves interfere with each other,
the interference will be completely constructive at a given time, but
then will slowly change into a completely destructive one as one
wave gets gradually out of phase with the other because of the different frequencies. The resulting superimposed vibration is also
shown in Figure 5.4. It has a frequency of 12 ½ per second, the
mean of the two superimposed vibrations. But the amplitude of
this resulting vibration is not constant, as was the case with the
other wave; it oscillates from large to small and back to large, and
the number of times it does that per second is equal to the *difference* of the two original frequencies. We then say that the amplitude is modulated by this difference frequency. As we know, a
change in amplitude means a change in loudness, and so the effect
of this superposition is a wobbly sound. If the difference in frequencies is one per second as in this example, the perception will
be a wobble. If the difference is, say, six per second, the sensation
no longer will be a wobble (since we cannot easily make out six
wobbles per second), but instead we will judge the sound unpleasant in a disquieting way. We will return to this phenomenon
when we discuss musical scales.

Pitch Perception

We started this chapter with a discussion of human loudness
perception, but changed to frequency perception when discussing
beats. Another important perceptional issue to be discussed,
which will be important for us in discussing music, concerns
human pitch perception. We already mentioned briefly that our
judgment of the pitch of a sound is connected with the frequency
of the sound. That is simple enough as long as the sound wave is of
a single frequency. Almost always, however, the sound we hear is a
mixture of many frequencies, or, to put it in a better way, the
sound wave has a shape that is very different from a simple sine
wave (see Chapter 2), and if we then Fourier-decompose that
waveshape into the usual set of sine waves, we get significant contributions from many such waves, and hence from many frequencies. When that is the case, which of those frequencies is registered
by our brain to be *the* pitch of sound?

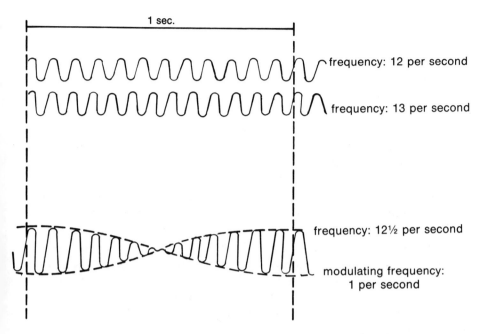

Figure 5.4 The formation of beats. Two waves with frequencies of 12 per second and 13 per second are superimposed and form an oscillation with a frequency of the average of the two former frequencies (that is, 12½ per second), the amplitude of which is "modulated" at a frequency that is the difference of the two former frequencies (that is, one per second in this example).

We know that in some cases (when we hear "noise"), the pitch cannot be determined at all. So let us restrict ourselves, for the moment, to "musical" notes for which we think we can determine a pitch. This pitch is *usually* the lowest frequency Fourier component that occurs in the sound we hear. If this were always the case, the matter would be simple and our brain would function like a physics instrument. Sometimes, however, the pitch at which our brain points does not correspond to the lowest Fourier component in the mix, but to a pitch lower than that, which *actually does not occur at all* in the mix. Under such circumstances our brain "imagines" a sound that does not exist, and identifies that as the pitch. We will see in Chapter 7 how this comes about. Even in advance of that discussion, however, it is perhaps appropriate to close this chapter with this "irrational" phenomenon, again underlining the fact that the human perception of sound is only partly a matter of what the ear registers, and the rest of it is all "in the head."

Summary

The perception of sound impressions by humans is a combined effect between the ear and the brain. We concern ourselves primarily with the former. In describing the functioning of the ear, we can, schematically, distinguish six components: the sound collector, the eardrum, the transmitter and amplifier, the linked window membrane, the cavity that contains the long membrane, and the tentacles with nerve endings.

Spatial localization of sound is accomplished with the help of our *two* ears, which utilize the following three effects: (1) The sound intensity at one ear is less than at the other because of the differing distances from the source and because the head "shadows" one ear. (2) There is a phase difference between the two waves reaching the two ears, due to the different distances from the source. (3) The arrival time of the sound is different at the two ears because of the difference in path lengths from the source. All these are more effective for high-frequency sound than for low-frequency sound.

We perceive loudness on the basis of the *ratios* (and not *differences*) in the amounts of energy per unit area falling on the eardrum. Hence the scale used to measure loudness uses such ratios. The ratio of the loudest to the softest sound we can perceive is 10^{14} or so in the sound energy per unit area, and this is said to correspond to 140 dB. Every *factor* of 10 in the sound energy per unit area corresponds to a *difference* of 10 dB, and every *factor* of two corresponds to 3 dB. Our loudness perception also depends on the frequency of the sound: We are most sensitive to about 4000 per second, and lose all perception at 20 or so per second at the low end, and at 30,000 per second or so at the high end.

Two sound waves at the same location that have a frequency difference that is very small compared with the frequency that either of the two produces, when superimposed, create a "beat" phenomenon. This is a sound with a frequency halfway between the two original frequencies, the loudness of which changes at a frequency that is the difference between the two original frequencies.

Human pitch perception can be curious. In some cases, for example, when we have a sound that is a mixture of several frequencies, our brain registers as the pitch of this sound a frequency that is lower than any that actually exists in the mixture.

Part III

Specifics of
Musical Sound

Chapter Six
Musical Sound

Rᴏᴜɢʜʟʏ sᴘᴇᴀᴋɪɴɢ, ᴡᴇ divide the sounds we hear into two categories: noise and music. This distinction is very rough indeed, since various criteria can be applied, which give different results. For example, we might say that a musical sound is one with which we can associate a pitch, whereas we cannot do that for noise. A different criterion is whether or not the sound is pleasing. (Think of the saying, "Music to my ear.") But even if there were only one criterion, the classification still might be difficult to make. Some noises have a discernible mixture of the pitchless and the pitched. Although the squeaking of a rusty door hinge is usually considered noise, one might be able to assign a pitch to the sound.

From the point of view of the physicist, a musical sound will be defined as a sound vibration most of which has a definite periodicity, and hence pitch. We may allow some admixture of components that are not periodic, but we will require that, to a large extent, the sound vibration, as plotted in time, should repeat over and over again. It will not repeat forever, of course, since sounds may last only a short time as measured on our usual time scale of a second, or half a second, or longer. Since we talk about audible sound, however, and since the human ear cannot perceive vibrations of which the frequency is lower than, say, 20 per second, we can say that even for sounds of very short duration, we can require that the vibration repeat itself 10, 100, 1000, or more times periodically, For sound of very low frequency and very short duration, this requirement may be violated, and indeed it will be difficult for us to tell, when we hear it, whether such a sound is music or noise. But in most of the cases that occur in everyday life, the foregoing definition is a useful one.

The analysis of musical sound then consists of plotting the sound-wave amplitude as a function of time. Such a plot will show a variation of the amplitude of sound on many different time scales: some features will manifest themselves on the time scale of 1/10,000 second, whereas for others it will be an hour or so. In the following these variations in sound amplitude are analyzed in terms of five time scales, or five time durations:

1. Supershort scale: 1/30–1/10,000 second.
2. Short scale: one second to 1/10 second.
3. Medium scale: one to five seconds.
4. Long scale: five seconds to one minute.
5. Very long scale: one minute to several hours.

As we will see, the complex and rich experience of music is made up of involved happenings on all of these time scales.

Supershort Scale

On this time scale, we want to consider one period of the sound vibration (you recall that if we talk about musical sound, such periodicity is in evidence), and investigate the waveshape within that period.

This shape will tend to be a wild one, very far from any particular sine wave (which, in musical language, is called a "pure tone"). To characterize this wild waveshape, we can determine its Fourier spectrum. Since the sound is periodic, its Fourier components will also be periodic, with frequencies that correspond to twice that period, to that period, to two-thirds of that period, and so on. Each of these components will be admixed in a strength indicated by the Fourier spectrum plot, as we saw in Chapter 2. Our ear receives the wild sound waveshape, and the brain registers it is an overall aural impression. The Fourier spectrum, therefore, is useful mainly for our analytical discussion.

How can we produce such a waveshape? We use sound sources, and combinations of these sources. These sources are the musical instruments, by themselves or in combination, including the human singing voice. More recently sources have also been produced by electronically driven (rather than humanly driven) instruments, but for our present discussion, this is an inessential difference.

Each musical instrument in itself produces a sound of which the Fourier spectrum consists of many frequencies in characteristic relative strengths. In an ensemble of instruments, such composite sounds are produced simultaneously. The lowest Fourier components (the "fundamental" frequency) of various musical "notes" played by various instruments will be different, and in music several such notes are usually sounded simultaneously. Our ear can usually decompose the overall sound wave into these several notes, expecially if we have had some practice.

Such a combination of notes can be pleasing to our ear (or, to be precise, to our brain), in which case we speak of consonance, or it can be unpleasant, or dissonant. The physical explanation for this will be discussed further in a later chapter. It should be added, however, that physics alone does not suffice. Combinations of notes judged consonant in one period of musical history (or one musical civilization) may be considered dissonant in another. In the past 300 years of Western music, the trend has been to become increasingly broadminded about what we call consonance.

Short Time Scale

This is the time scale (up to a second or so) in which we can describe individual musical notes in terms of their time evolution over tens, hundreds, or thousands of periods. The duration of a single note in music tends to be of this time scale, from a tenth of a second or so to one second. During this time the amplitudes of the note and its Fourier components do not remain the same. The overall sound energy produced by the note will tend to change, and in addition even the relative strengths of the various Fourier components as compared with each other may change. A note sounded on a piano, for example, starts rather suddenly (that is, the overall sound energy of the note rises from zero to some substantial value in a time duration corresponding to a tenth of a second or less), but then it continues to sound, with rapidly decreasing strength, for a second or more. It can be shown that during this time even its Fourier spectrum changes, since some Fourier components die out more quickly than others.

This time evolution of the overall amplitude of the note plays a very important role in our differentiating among different instruments. Tape some music being played on a piano and then play the tape backward. The result will sound ludicrous and altogether different from that played on the piano.

In musical language we describe the various qualities of sound different instruments can produce as different timbres. Although they may be playing the same musical note (that is, the fundamental frequency is the same), the sound of a piano will be radically different from that of an oboe, and again different from that of a singing voice. Even an ear completely untrained in music will note the difference. This difference is attributable to two main factors: the different Fourier spectrum, and the different time evolution of the sound of this second, short time scale.

And the same musical note, played on the same instrument, can still exhibit variations, depending on the overall loudness of the sound. We can touch a guitar string gently or we can pluck it fiercely, thus producing quite different sounds. The difference arises not only from the different Fourier spectrum and the different time evolution on the second, short scale, but also from the difference in overall loudness. In musical language the dimension of music that concerns loudness and softness is called dynamics.

Medium Time Scale

We are now up to a time duration of a few seconds. In most music a given note generally is not sounded for more than a second, and so on this time scale we are confronted with the transition from one note to another as time goes on. The relative time durations of such notes determine what in musical language is called rhythm. Together with consonance (and dissonance), rhythm is one of the two basic elements of any music. Our body, through its various periodic functions (such as heartbeat, breathing, wake-and-sleep cycles, life and death, etc.) has its own rhythms, and so music, which also has a rhythm, seems natural to us. Innumerable rhythmic patterns appear in various types of music, and even a given piece of music can simultaneously feature several rhythmic patterns, sounded by different instruments.

Long Time Scale

We have now reached the range of many seconds, or a minute. During this time interval, music is divided into musical phrases, just as our speech is divided into phrases or sentences. In music, as in speech, the end of a phrase is usually marked by a small pause, or at least a change of pace in rhythm and note structure.

Very Long Time Scale

We refer to musical pieces or musical compositions. These may be songs; they may be written for one instrument alone; they may be for a large ensemble of instruments (orchestra); or they may be for orchestra and voices together. In any case, after a considerable time duration of more or less continuous music, the performance comes to a halt, which marks the end of that piece. Some forms of music also have intermediate stops. For example, a Western classical musical form called the symphony usually has several movements, after each of which a minute or so of rest is taken. Similarly, a Western musical form of singing and orchestral playing called opera has, just as do theatrical presentations, several acts, between which there are 15–20-minute intermissions to give the musicians and audience a rest. In the music of India, various *ragas* are used in the same musical performance, which somewhat demarcate the overall sequence of the music. The same is true of jazz concerts, or the art of popular songs.

We expect that one such long unit of music, separated from the next by a considerable pause, should have its own logical structure, its own coherence, its own pattern. A short song, for example, has a topic and a message or a mood that binds it together. In a movement of a classical symphony, musical themes (analogous to phrases on the fourth time scale) appear and reappear, the dynamics slowly builds up and then subsides, the harmonies change from one musical key to another, and so forth. This large-scale structure of music is often a good test of the greatness of the composer who created the music; to build such a convincing structure, the composer must not only be skilled in juxtaposing notes in a pleasing way, or in forming interesting rhythmic patterns, but also must "have something to say musically" on a much longer time scale. The difference also has its analogy in the spoken word. There are people who are clever with words and can put together attractive rhymes, but to be great writers they also have to have something to say; they must have an extended, long-range subject for communication that impacts on the audience or readership, penetrates deeply, and sustains interest.

Thus music consists of elements and patterns superimposed on each other on a number of quite different time scales. This structure of "nested" time scales is also characteristic of many other phenomena in the world. The weather, for example, which can be characterized by such quantities as temperature, amount of sunshine, humidity, and wind velocity, changes periodically ever day, between day and night. On a somewhat longer time scale, sunny days alternate with rainy ones, warm ones with colder ones. Then, on a yet longer time scale, the seasons enter and summers alternate with winters. To go beyond that, in the history of the earth, glacial periods that lasted thousands of years have alternated with equally long time intervals when the temperature was unusually high and ice almost completely vanished.

The Stereo Effect

So far we have discussed only the *time evolution* of the sound reaching us in music. As we know, however, from Chapter 5, we can also specify the spatial location of the source of sound reaching us, and hence another dimension of music opens up; this is the stereo effect. For example, if the music is created by a large assortment of musical instruments dispersed over a large area, our ear can locate the various instruments in space even if they all play simultaneously. Indeed, real-life experiences with such ensembles of instruments have made us used to hearing such a stereo effect, and so we cherish it even in situations where recorded music is performed in a very different environment and the effect must be produced artificially. Composers of music are well aware of this additional dimension, when they prescribe offstage fragments of music in opera performances, or distribute musical material among different instruments with quite different locations on the stage of the concert hall. One of the most monumental pieces of music ever written, the "Requiem" of the French 19th-century composer Hector Berlioz, which requires several hundred people (singers and instrumentalists) to perform, achieves one of its unique effects through separate ensembles of brass instruments located in the back of the concert hall, in the midst of the audience, and high up, to play together with the others on the stage.

Most of the physics we will discuss in connection with music pertain to the happenings on the two shortest time scales, namely, the supershort scale (a small fraction of a second) and the short scale (up to a second). The other three scales involve aspects of music that are, in our present state of knowledge, better treated in the usual music courses, using the language of musical analysis itself—that of esthetics, of human perception, of psychology. This does not mean we can be sure that physics has no relationship to musical esthetics or to the psychology of music. On the contrary, we are almost certain that eventually the connection will be made, not by proving that esthetics is nothing but physics, but by realizing that both physics and psychology have their own languages to describe the same set of phenomena, and that we can explain a certain aspect of music in many different ways. Such an advanced stage of understanding is, however, still far in the future, and so we find physics much more suitable for discussing the first two time scales, and unsuitable for dealing with the longer range structure of music.

Summary

From our point of view, the difference between noise and music
is that the latter consists mostly of sound vibrations that have a
definite periodicity, and hence pitch. We can analyze musical
sound on five different time scales. The *supershort scale*, which
lasts 1/30–1/10,000 second, corresponds to one periodicity of the
sound wave. The shape of this wave within one period will be
characteristic of the source (instrument) that produced it. Some
combinations of notes are pleasing to our ears and some are not,
thus defining consonance and dissonance, but the division between the two also depends on cultural conventions. The *short time
scale*, up to a second, corresponds to tens, hundreds, or thousands
of periods. The evolution of sound in terms of its Fourier spectrum
and volume contributes greatly to the different characters of musical instruments. The third scale, called *medium scale*, up to a few
seconds, defines rhythm in music. The fourth, or *long time scale*,
up to many seconds or a minute or so, consists of musical phrases.
Finally, the *very long time scale* is characteristic of entire musical
pieces, compositions, or subdivisions of such compositions, and
lasts many minutes or a few hours. This nested set of time scales
resembles the rhythms we find in other facets of life, also.

Spatial distribution of sound contributes to musical effects,
such as the stereo effect that envelops the listener. Scientific
analysis of music pertains mainly to the first two time scales,
whereas musicological and esthetic terminology are used in describing the longer time scales. These are complementary ways of
talking about music.

Chapter Seven
Pitch and Musical Scales

As we saw, a musical sound has a given fundamental frequency, and mixed with it higher frequencies coming from the Fourier components of the decomposition of the actual waveshape. In other words, the musical sound is a periodic vibration, and this period determines its fundamental frequency. But the shape of the sound wave is virtually never that of a simple sine wave, and so if we want to describe it, in the Fourier manner, as a superposition of sine waves, we have to add to the sine wave of the fundamental frequency sine waves with twice, three times, and so on, that frequency. These higher frequency components are called the first, second, third harmonics or overtones.

It might be worthwhile here to review briefly the terminology of harmonics, overtones, and "partials," which is often confusing and inconsistent.

Overtone is any higher frequency vibrational mode of a system, excluding the lowest frequency mode, called the fundamental. If the frequencies of these overtones are exactly or approximately integer multiples of the fundamental, then they are also called harmonics. Usually, but not always, the fundamental is not counted as a harmonic, so that the frequency of the first harmonic is already some nonunity integer times the fundamental frequency.

The term "partials" is used for overtones and the fundamental, regardless of whether or not they are harmonics.

In this discussion the term partial will not be used. Since in most cases the overtones we will consider will also be harmonics (exception: percussion instruments), the two terms will be used interchangeably, but unambiguously. The fundamental will not be called first harmonic but rather the one with the next lowest frequency will be designated the first harmonic.

The frequencies of these overtones are then in the ratio 1:2:3:4.... The brain defines the pitch of a sound to be that of the fundamental frequency. This is true in almost all situations. But there are ways to "fool" the brain, as alluded to in Chapter 5.

Consider a sound with the above series of overtones, from which we *remove* the fundamental component altogether. This can be done electronically, by placing frequency filters in the path of the sound, which can remove certain frequencies. The resulting sound will have a series of components, starting with the former first harmonic, and so the frequency ratios now will be 2:3:4.... This is *not* the sequence our brain is usually accustomed to hearing, since, for example, the ratio of the two lowest frequencies now is 1.5 and not 2, as it was before. Under such circumstances our brain tends to register the sound *as if* it contained the missing former fundamental frequency also, and so it will assign to the sound the pitch this missing fundamental has, even though that frequency is not present in the sound our ear receives.

The ensemble of higher frequencies the ear receives in the higher Fourier components of the sound is registered in the brain as the timbre, the quality of the sound, as we saw in the previous chapter. Indeed, such overtones greatly contribute to the beauty of the sound we hear. A pure tone with no overtones sounds very dull and uninteresting.

In what follows, we discuss the sets of musical notes (each determined by its fundamental frequency) used in creating music.

The Simple Fractions Requirement

In principle, we could use notes of any frequencies in a musical composition. In practice, however, we find it useful and pleasing to form "scales" of notes, each containing a finite number of different frequencies, and we then utilize, in a particular musical composition, mostly the notes of one such scale. The reasons for this are numerous. For one thing, if we are to sound several notes simultaneously, we must assure that they form a consonance; in other words, they sound pleasing when played together. Unless the frequencies of the various notes in the scale are chosen carefully, sounding two notes can cause dissonances. Consider, for example, two notes, one with a frequency of 440 per second and the other with a frequency of 222 per second. The first overtone of the latter would have a frequency of 444 per second, and that overtone, heard with the first of our notes (at 440 per second), would create a beat with a wobble of four per second, and thus cause an unpleasant sensation.

How can we avoid this? It is not difficult to see that if the *ratios* of the frequencies of two notes in the scale can be expressed as a fraction with a small numerator and a small denominator, we can avoid such *near* coincidences between the frequencies of overtones of two different notes. This is illustrated in Table 7.1 for the case when the ratio is 5/4, and the fundamental frequency of the lower of the two notes is 400 per second. In this table no two frequencies differ from each other by just a small amount so that beats could develop.

Thus our first criterion for including musical notes in a scale is that they should have frequency ratios that can be expressed as fractions with small numerators and small denominators. We call this the "simple fraction requirement."

Building the Scale

We now use this to build a scale. In various types of music around the world and throughout history, many kinds of scales have been used. One of the most common ones in Western music in the last few hundred years, for example, is called the major scale.

In building such a scale, the first decision to make is to have some anchor points in the frequency range. For music we need to define a scale only within the audible range, that is, from, say, 20 per second to 5000 per second. But even this range is quite broad,

Table 7.1

Frequencies of Fundamentals and Overtones of Two Musical Notes, at 400 per Second and at 500 per Second

400	800	1200	1600	2000	2400	2800	3200	etc.
500	1000	1500	2000	2500	3000	3500	4000	etc.

and unless we want to space the notes far apart, we have to specify quite a few notes within that range. Instead we decide that our scale will be periodic in frequency, that is, we will have the same kind of a set of notes repeated over and over again, always located between two anchor frequencies. We will use as such anchor points a set of frequencies in which two neighboring ones have a frequency ratio of 2:1. Two such notes will be called to form an oc-tave. (One might wonder why a ratio of two is denoted by a word suggesting eight, or "octo" in Latin. The answer will be given shortly.)

Note that these octaves are defined not by frequency *dif-ferences* but by frequency *ratios.* If, for example, we start with a note of frequency 440 per second (known in the musical world as a standard A), the octaves above this note will have frequencies of 880, 1760, 3520, and so forth, and below it of 220, 110, 55, and so forth. Two notes an octave apart, when played together, sound pleasant (though not particularly interesting).

Now we will place additional notes between these anchor points on the frequency scale, and the frequencies of these additional notes will be given as *ratios relative to the* frequency of the lower of the two notes in the octave. If these ratios satisfy the simple frac-tion criterion, the ratios with notes in other octaves will auto-matically do so as well.

The next decision to make concerns the number of notes we want to cram into an octave. Opinions on this differ. In the case of the major scale that we use as an example, the number chosen is eight, including the two notes that form the bounding octave. (This is the reason why the name of an octave refers to eight.)

We have to make one last decision before our scale is completely set: What ratios should we use for the frequencies of the notes? For the moment we again will use the guideline of the simple fraction criterion, and thus choose a sequence of simple fractions, which, after seven steps, gets us to a total ratio of two.

In the so-called diatonic major scale, this is done by choosing the ratios of six intermediate notes with respect to the bottom note in the octave to be (in order of increasing frequency) 9/8, 5/4, 4/3, 3/2, 5/3, and 15/8. Let us call the eight notes in our scale by Roman numerals: I, II, III, IV, V, VI, VII, VIII. We see then, from Table 7.2, that not only are the ratios of frequencies given by simple fractions with respect to one, but the ratio of *any* two notes is expressed by a reasonably simple fraction.

Note, however, that 9/8 and 15/8 are not as simple as 4/3 or 3/2. Indeed, sounding the first two notes on our scale together, or the first and the seventh together, produces a much less smooth sensation as sounding the first and the third, or the first and the fifth.

We see that the notes within an octave in the diatonic major scale are not situated so that the ratio of two neighboring notes is always the same. These ratios are 9/8, 10/9, 16/15, 9/8, 10/9, 9/8, 16/15. The 9/8 and 10/9 are not too different, but the 16/15 is a much smaller ratio than 9/8 or 10/9. Thus in the ascending scale we have steps that are, starting from the lower anchor point, large, large, small, large, large, large, and small, in this order.

The Equal Footing Requirement

So far, so good. But now we arrive at a second, equally reasonable demand for the construction of scales; namely, that we should be able to get the same scale no matter on which note we start. In other words, we want to be able to build a scale using as an anchor point any of the notes we have defined.

In particular, let us say that we have taken as the lower anchor point a frequency of 440 per second, and built a diatonic major scale on it. The frequencies of the notes in the first octave above this 440 anchor note will then be 495, 550, 586.67, 660, 733.33, 825, and 880. Now take any of these notes, say, the one at 586.67, and with that as a new anchor point, build a diatonic major scale *on it,* using the universal ratios that hold for *any* diatonic major

Table 7.2
Ratios of Frequencies of Notes in a Diatonic Major Scale

	I	II	III	IV	V	VI	VII	VIII
I	1	9/8	5/4	4/3	3/2	5/3	15/8	2
II	8/9	1	10/9	32/27	4/3	40/27	5/3	16/9
III	4/5	9/10	1	16/15	6/5	4/3	3/2	8/5
IV	3/4	27/32	15/16	1	9/8	5/4	45/32	3/2
V	2/3	3/4	5/6	8/9	1	10/9	5/4	4/3
VI	3/5	27/40	3/4	4/5	9/10	1	9/8	6/5
VII	8/15	3/5	2/3	32/45	4/5	8/9	1	16/15
VIII	1/2	9/16	5/8	2/3	3/4	5/6	15/16	1

scale, no matter where it begins—namely, the 9/8, 5/4, 4/3, and so on listed before. In doing so, of course, one hopes to be able to use the notes we arrived at in constructing our first diatonic scale (anchored at 440) and so not have to introduce many new notes.

The result is chaos. The process is shown in Table 7.3. We see from that table that by the time we have built a diatonic major scale on each of the seven notes of the original diatonic scale, we have defined 19 notes, including the two anchor points of the octave. And, that is not the end of it; if we wanted to be consistent, we would now have to build diatonic scales on each of the 11 new notes also, thus undoubtedly creating more notes in the process. Soon the interval within the octave would be jammed with dozens of notes.

One might ask what difference that makes. The main difficulty with such an army of notes arises with regard to musical instruments on which the pitches of playable notes are fixed. (Not all commonly used instruments are in that category; for example, those in the violin family are not. But most woodwinds are, as is the piano.)

We have arrived at a crisis: *The two eminently reasonable requirements*—namely, the simple fraction criterion (to assure consonance) and the requirement that all notes should be on an equal footing from the point of view of being able to build a diatonic scale on them (which we call the equal footing criterion)—*are incompatible* with each other. We cannot satisfy both of them simultaneously, and so must compromise.

Table 7.3
Generation of the "Complete" Diatonic Major Scale

1	$\frac{9}{8}$	$\frac{5}{4}$	$\frac{4}{3}$	$\frac{3}{2}$	$\frac{5}{3}$	$\frac{15}{8}$	2						
$\left(\frac{135}{128}\right)$	$\frac{9}{8}$	$\frac{81}{64}$	$\frac{45}{32}$	$\frac{3}{2}$	$\frac{27}{16}$	$\frac{15}{8}$	$\frac{135}{64}$	$\frac{9}{4}$					
$\left(\frac{25}{24}\right)$	$\left(\frac{75}{64}\right)$	$\frac{5}{4}$	$\frac{45}{32}$	$\frac{25}{16}$	$\frac{5}{3}$	$\frac{15}{8}$	$\frac{25}{12}$	$\frac{75}{32}$	$\frac{5}{2}$				
(1)	$\left(\frac{10}{9}\right)$	$\left(\frac{5}{4}\right)$	$\frac{4}{3}$	$\frac{3}{2}$	$\frac{5}{3}$	$\frac{16}{9}$	2	$\frac{20}{9}$	$\frac{5}{2}$	$\frac{8}{3}$			
(1)	$\left(\frac{9}{8}\right)$	$\left(\frac{5}{4}\right)$	$\left(\frac{45}{32}\right)$	$\frac{3}{2}$	$\frac{27}{16}$	$\frac{15}{8}$	2	$\frac{9}{4}$	$\frac{5}{2}$	$\frac{45}{16}$	3		
$\left(\frac{25}{24}\right)$	$\left(\frac{10}{9}\right)$	$\left(\frac{5}{4}\right)$	$\left(\frac{25}{18}\right)$	$\left(\frac{25}{16}\right)$	$\frac{5}{3}$	$\frac{15}{8}$	$\frac{25}{12}$	$\frac{20}{9}$	$\frac{5}{2}$	$\frac{25}{9}$	$\frac{25}{8}$	$\frac{10}{3}$	
$\left(\frac{135}{128}\right)$	$\left(\frac{75}{64}\right)$	$\left(\frac{5}{4}\right)$	$\left(\frac{45}{32}\right)$	$\left(\frac{25}{16}\right)$	$\left(\frac{225}{128}\right)$	$\frac{15}{8}$	$\frac{135}{64}$	$\frac{75}{32}$	$\frac{5}{2}$	$\frac{45}{16}$	$\frac{25}{8}$	$\frac{225}{64}$	$\frac{15}{4}$

The top line gives the ratios of frequencies in the diatonic major scale. Those ratios constitute the definition of this scale. The second row gives the ratios of a similar scale if we start on the second note of the previous line, that is, the numbers of the second row are the numbers of the first row multiplied by 9/8. The third row gives the ratios of a similar scale if we start on the third note of the first row, that is, the numbers in the third row are the numbers of the first row multiplied by 5/4. Etc. The numbers in parenthesis are the ratios (transposed down by an octave) which fall above the upper end of the octave of the first row, and are obtained by dividing those ratios by 2. (E.g. in the second row, 135/64 divided by 2 gives 135/128.) One can see that within an octave (i.e. between the ratios of 1 and 2) there are 19 different ratios.

The Compromise

To see how a good compromise can be reached, let us plot the various frequencies we get by building all those diatonic scales. In Figure 7.1 the frequencies for the seven diatonic scales composed in Table 7.3 are plotted. On that figure we identified, as an example, the previous notes of I, II, and so on with the notes that, in musical notation, are called C, D, E, F, G, A, B, and C again. This is termed the C-major diatonic scale, since it starts on a note called C. This change of notation is not essential to our arguments, but it does bring us closer to the notation commonly used in music.

We see from Figure 7.1 that the 19 frequencies can be grouped into 12 (not counting the upper C) by combining certain pairs of frequencies that are fairly close together. Figure 7.1 does not include all those notes that we get by also building diatonic scales on top of the 11 new notes (as mentioned earlier), but it turns out that even those notes can be reasonably accommodated within this 12-note scale. To look at it in another way, what we did was to insert an extra note between those two notes of the original diatonic scale that were a "large step" apart.

We see from Figure 7.1 that, as drawn there, the 12 groups of notes are not necessarily regularly distributed along the frequency scale. If we look at the ratios of the frequencies of two neighboring groups, we find somewhat larger and somewhat smaller values.

Since we are in the process of compromising, by combining notes into an "average" frequency for that group, we might as well choose a compromise that is easy to use and that at least satisfies the spirit of one of our two requirements—the equal footing criterion. Accordingly we will choose the 12 steps between the lower and upper C's so that the *ratio of two neighboring frequencies is always the same.*

We have 12 steps in the octave, after which we must reach an overall ratio of two. Thus by applying the ratio of neighboring frequencies 12 times in succession, we must get two. "In succession" means that the ratios of the 12 steps are multiplied by each other. Thus we need a ratio for the two neighboring steps, which, raised to the 12th power, gives two. Such a ratio is the 12th root of two, denoted as $\sqrt[12]{2}$. If this seems complicated, simply remember it as a number that, when multiplied by itself 12 times, gives two. The numerical value of that number turns out to be 1.059463 . . . , an infinite decimal fraction. Each such step is called a *tempered semitone* step.

Thus we have created another scale, which we call the tempered scale. It is different from the diatonic scale in that the ratios of frequencies of the notes in it compared with the lower end of the octave are different. As we constructed it, this new scale has 12 notes, but if we take out the second, fourth, seventh, ninth, and 11th notes, we are left with seven notes with frequency ratios *approximately* the same as the frequency ratios of the seven notes in the diatonic scale. This is shown in Figure 7.2.

In our tempered scale, two notes sounded together will not be as pleasing as they were in the diatonic scale. For example, consider the two scales, both built on the lower anchor point of 440 per second. In the diatonic scale, there will be a note with a frequency of 1.5 times 440, or 660 per second. The corresponding

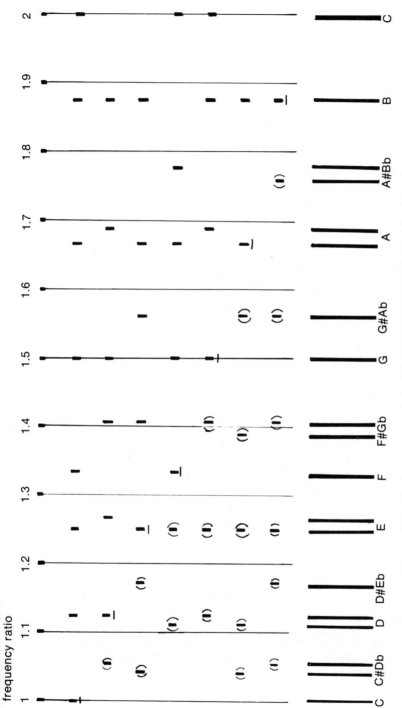

Figure 7.1 The frequency ratios of Table 7.3 and the musical notation of the notes that are associated with those frequencies if the ratio 1 is taken to be the musical note C.

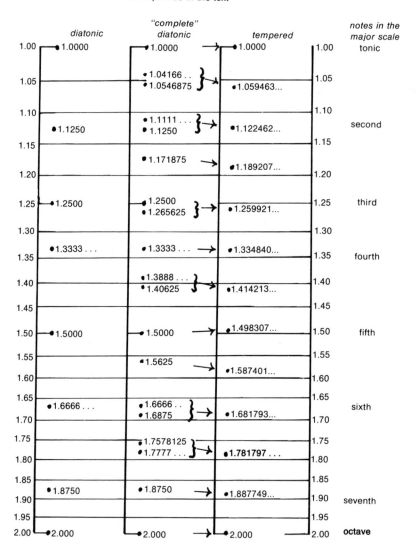

Frequency ratios for diatonic and tempered scales
as explained in the text

Figure 7.2 The frequency ratios in the diatonic major scale, in the "complete" ("chromatic") diatonic scale, and in the tempered scale, together with the names of the musical intervals. The ratios indicated in the middle column for the complete diatonic scale are those shown in Figure 7.1 also.

note in the tempered scale will have a frequency of 440 times 1.4983 . . . , or 659.25 . . . per second. The second overtone of the 440 will be at 1320 per second, while the first overtone of the 659.25 . . . will be at 1318.50 . . . per second. The two together will produce a beat with a wobble of one every two seconds, and even if this wobble is not conspicious, it creates a subconscious uneasiness in the listener. In the diatonic scale, both frequencies would be 1320, and hence no beats would develop.

On the other hand, in using the tempered scale, the sequence of notes (i.e., a scale) started on any note will have the same frequency ratios to each other as a similar sequence started on any other note. We have achieved equal footing, and had to insert, between two anchor frequencies, only 11 intermediate notes. This allows us to build fixed-note instruments in a practicable way. For instance, on a piano, between two C keys there are six white keys and five black keys, giving the 12 tempered semitone intervals between one C and the next. The tempered C-major scale will involve all the white keys and none of the blacks. We could, however, just as well use as anchor points two G keys on the piano. In this case, a tempered scale, with exactly the same frequency ratios as before, can be played on the existing keys of the piano, using, as it turns out, six white keys and one black key.

There are many other scales in addition the diatonic and tempered scales. Even in western music, the Pythagorian scale played an important role in Renaissance music: the pentatonic scale is used by Bartok. In the music of cultures other than the West, such as in India, more elaborate scales are standard. In some types of contemporary Western music, extensive experimentation is under way with new scales. The aim of this chapter, therefore, was not to offer an encyclopedic discussion of scales, but to illustrate the *principles* that underlie scale formation, as a basis for dealing with any of the other scales in use.

Summary

We constructed a sequence of musical notes (each with a definite fundamental frequency) and we called it a scale. To do so we first divided the audible range into octaves, an octave being bounded by two notes with a frequency ratio of two. In each octave we then inserted six intermediate notes of fixed frequency ratios with respect to each other. For a pleasing sound impression, we want these ratios to be fractions with small numerators and denominators (the "simple fraction" criterion). With ratios selected (with respect to the lower end of the octave) as 9/8, 5/4, 4/3, 3/2, 5/3, 15/8, we obtain a diatonic scale.

We then try to impose a second requirement, namely, that one should be able to use the same frequency ratios to build a diatonic scale on any of the notes we have in our first diatonic scale (not only on the two anchor notes), without having to add too many new notes into a given octave. This turns out to be impossible. As we try it, we accumulate a huge number of additional notes in our octave. It is clear that this second requirement (the equal footing criterion) and the simple fraction criterion are incompatible.

We then strike a compromise by modifying slightly the frequencies that would be given by the simple fractions, and also yield on the number of notes between the two anchor notes, raising it from six to 11. We thus obtain a different scale, in which the two neighboring notes within the octave always have a frequency ratio of $\sqrt[12]{2}$. This tempered scale is not quite as mellifluous as the diatonic was, but it achieves the goal of putting all notes on an equal footing in that a scale built on any note will sound relatively the same as a scale built on any other note. This allows the construction of practicable fixed-toned instruments, such as the piano.

Part **IV**
Making Music

Chapter Eight
Musical Instruments

Iₙ ₜₕₑ ₙₑₓₜ five chapters, we discuss the physics of musical instruments—that is, of gadgets used to produce musical sound. In various cultures throughout the world, and in various ages in human history, many hundreds of different instruments have been constructed and used. They may look very different from each other, and yet all have certain basic features in common. The purpose of this chapter is to discuss these universal features of instruments, so that we can apply them in our subsequent discussion of specific instruments.

The Main Components

There are five main components that can be identified in virtually all musical instruments. They are easy to pinpoint in instruments that are directly activated by a human being, and such instruments make up the majority of instruments in existence now and in the past. The more recent "electronic" instruments, in which the sound is not produced directly by the vibration of a mechanical object but by the oscillation of electric and magnetic fields in an electronic circuit, also include these main components, although it might be more difficult to identify them without knowing something about electronics. In any case, however, our attention will be directed mainly to nonelectronic, "old-fashioned" instruments, in which these components are clearly evident.

The five components are: energy source, the energy transmission from the source to the instrument, the primary vibrator, the resonant vibrator, and the sound effuser, which directs the sound out of the instrument toward the listener.

Energy Source

Since sound waves represent energy, and since we know that in any closed system the total amount of energy must be conserved, when sound waves are generated some source must pour energy into the system that produces them. In humanly activated instruments (which include all the usual instruments except the electric organ, the motor-driven hurdy-girdy, and electronic sound generators) the energy comes from the human body, in one form or another. For example, in the case of the violin, the muscle movements of the arm holding the bow are the source of the energy. For a drum it is the muscle movements of the arm holding the drumstick. For a tuba the energy comes from the muscles contracting the lung. In the case of a piano, the energy resides in the movements of the player's fingers hitting the keys.

The amount of energy expended by somebody playing a musical instrument is not inconsiderable. Energy expenditure per unit time is measured in units of a watt. If you lift a pound to a height of three feet in one second, you expend about five watts. The amount of light and heat energy emitted by a 100-watt light bulb per second is, of course, 100 watts. One can estimate that a person playing an instrument such as a violin, a trumpet, or a kettle drum expends about 50 watts. It is most interesting how *little* of this energy ends up as sound wave energy. The efficiency (that is, the fraction of the energy originally expended that appears in the sound waves themselves) is about 1 percent. The rest turns into heat, following the second law of thermodynamics, but never passes through a phase of sound wave energy. The energy in sound waves, in an auditorium, for example, also ends up as heat after the waves collide and are absorbed by the walls, one's clothing, and so forth.

Considering the tiny amount of energy generated by the sound waves thus produced, it is remarkable that we can hear an orchestra at all. The total sound power output of a large orchestra is less than the power output of a single 100-watt light bulb. What permits us to enjoy the music is the equally remarkable sensitivity of our ears. An auditory sensation of 60 dB corresponds to an actual energy per second falling on your eardrums of only about 10^{-10} watts. So our ear can overcome the great inefficiency of the sound-production mechanisms by being supersensitive to the trickle of sound energy that is produced.

Energy Transmission

When your muscles expend energy, there must be an energy transmission mechanism to channel that energy into or onto some part of the instrument in order to generate some vibration of that part. In the case of a drum, this is done with the drumstick, which carries the energy in your swinging arm onto the surface of the drum. In the case of the piano, the energy expended by your fingers is conducted through the piano keys and a series of levers linked with the keys to the piano strings inside the piano body. In the case of a wind instrument (a flute, clarinet, trumpet, tuba, horn, etc.), the carrier of energy is an airstream you blow out of your lungs and into or onto the instrument. For a violin the transfer mechanism is the bow. It has a group of hairs treated with a special material so that when you draw it across the strings of the violin, the bow alternately grabs and then releases each string, thus making it vibrate. In the case of the harp or the guitar, the transfer is more direct: Your fingers pluck the string, that is, pull it out of its equilibrium position, and hence initiate a vibration. When you sing your transfer mechanism is similar to that of a wind instrument. The air you exhale from your lungs moves fast, and thus has considerable kinetic energy; this air impinges on your vocal cords and makes them vibrate.

The Primary Vibrator

The energy transferred to the instrument sets up vibrations in some part of the instrument designed for this purpose. It is at this stage that the human energy originally invested in instrument playing turns into a potential of producing sound, since vibrations (in the right frequency range) can then cause waves in a transmitting medium. As we will see, however, this primary vibration directly induced by the energy fed into the instrument neither is strong enough nor has the right Fourier spectrum to produce beautiful sound.

In various types of instruments, the primary vibrator has very different forms. On string instruments such as the violin, guitar, double bass, and dulcimer, it is a string. In the so-called woodwind instruments, the primary vibrator is a reed or a pair of reeds. In brass instruments it is the player's lips, and in a flute it is the air itself that is close to the edge of the hole onto which the player blows. In the various drumlike instruments, the primary vibrator is an elastic membrane. In a piano, although this is also a percussion instrument, the primary vibrator is a string, as in the string instruments. We see that, except in the flute, the primary vibrator is some material object with some elasticity in which the energy transfer excites vibrations.

The Resonant Vibrator

To understand the function and the workings of this element of instruments, we have to understand the concept of resonance. It is a concept familiar from various everyday experiences, to which I will refer.

There are two parts to our understanding of resonance.

CHARACTERISTIC FREQUENCIES
OF A VIBRATING SYSTEM

First, most systems that are able to vibrate at all have certain "favorite" frequencies in which they vibrate, and in which they will vibrate if you leave them to their own devices. Take, for example, a string of a given length, and suspend on it a weight large enough to keep the string taut. Then give the weight a nudge so that it can swing back and forth like a pendulum. You will find that this simple pendulum always will swing with the same period (or same frequency), which will be independent of the size of the weight, and of whether you give the weight a small initial nudge or a heftier one. The period will depend on the length of the string, but if you do your experiments with the same length of string, your system will exhibit a great preference for this particular frequency.

You can repeat the same experiment with a weight at the end of a spring, in which case you can initially stretch the spring and then watch the weight jump up and down. Again, for a given spring, you will find a characteristic frequency, independent of the size of the weight, and independent of the size of the initial stretch, though dependent on the strength of the spring. But for a given spring, you will have such a characteristic frequency.

As a third example, consider that your car is stuck in mud, and you try to push it out. You strain and manage to get the car wheel away from the lowest point of the rut. Then you get tired, let the car go, and it will roll back, overshoot, roll forward again, overshoot again the lowest position in the rut, then roll back again, and so on. No matter how you manage to get the car swinging in the first place, it will have a characteristic frequency of rolling back and forth in the rut.

For another example consider the sun, which has certain spots whose number varies with time. People have found, by observation, that the maximum and minimum numbers of sunspots alternate fairly regularly, and that this oscillation has a frequency of about 11 years. We do not understand why this is so, but we know that the very complex mechanism inside the sun that causes these spots must be such as to have a characteristic fluctuating frequency of 11 years. Knowing this will help us determine what this mechanism may be.

With regard to music, if we consider a vibrating string, we will also see that the vibrations of such a string will exhibit some characteristic frequencies. So will the wooden sheet that forms the top or the bottom of a violin, or the column of air inside the tube of a trumpet, or the large wooden plank that is built into a piano as a "sounding board." In all these cases, more than one characteristic frequency is associated with the same object. For example, on a string that is clamped down at both ends, we can have a standing wave (which is a vibration of the string) with no nodes other than at the two ends, or a standing wave with an extra node halfway between the two ends, or one with two nodes, and so on. Each of these vibrations will have a different characteristic frequency. In the case of two-dimensional objects of odd shapes (such as the top of a violin), there is an even greater variety of characteristic vibrations, and hence characteristic frequencies, as indicated in later chapters.

Thus we can see that each system that is capable of vibrations has its own characteristic vibration patterns, and hence its characteristic frequencies. These characteristic frequencies are also referred to as *resonant* frequencies, for reasons that will be evident presently.

BEING IN RESONANCE WITH A SYSTEM

Now we turn to the second part of our understanding of what a resonance is by asking ourselves how we should feed energy into a system to get it to vibrate. Consider, as an example, a playground swing, with a child sitting on it. Initially the swing is at rest, but you want to get it moving so the child can have fun. You push the swing, and it starts swinging up, then stops, and starts swinging back. To get it to swing more vigorously, you wait until the swing has overshot, swung in the other direction, stopped, and then started swinging again in the original direction, and you push again. In this way each push works *with* the "natural" motion of the swing, and hence adds to it. In our fairly newly acquired terminology, we can say that your efforts and the natural oscillation of the swing interfere constructively.

Imagine that the characteristic frequency of the swing is six seconds. You then do best by pushing on the swing (always in the same direction) every six seconds, thus always being in phase with the swing. If, instead, you decide to push on the swing every 2.4 seconds, you sometimes will help the natural motion of the swing, but at other times will work against it, and so, on the average, will accomplish nothing. This is shown in Figure 8.1.

If you always reinforce the motion of a vibrator in the form of such a constructive interference, you are said to be *in resonance with* the system. You are in resonance with the system because in this way you can excite the resonant frequencies of that system, in other words, transfer energy to the system so as to excite its characteristic vibrations. We see that to be in resonance with a system, you should have the same frequency as one of the resonant frequencies of the system.

If energy is carried to a system by another system with many characteristic frequencies of its own, a resonance will be established between the two systems if one or several of the characteristic frequencies of the second system coincide with one or several characteristic frequencies of the first. In that case energy can be easily transferred from the first system to the second. If that is not the case, energy will not be transferred to the second system even if an opportunity is offered, in principle, for such an energy

time (seconds)

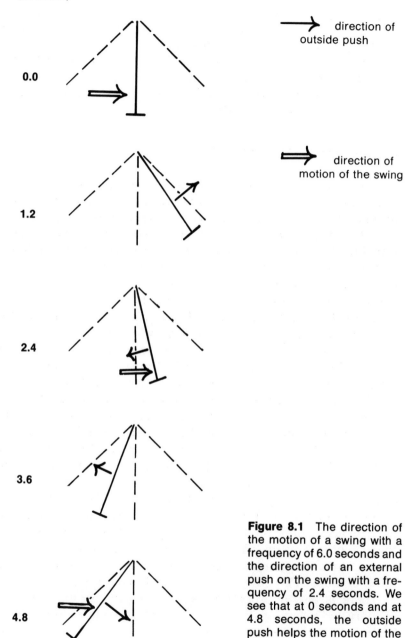

direction of
outside push

direction of
motion of the swing

Figure 8.1 The direction of
the motion of a swing with a
frequency of 6.0 seconds and
the direction of an external
push on the swing with a frequency of 2.4 seconds. We
see that at 0 seconds and at
4.8 seconds, the outside
push helps the motion of the
swing, but at 2.4 seconds it
impedes it.

transfer by linking the two systems in some way. In our previous example, if you shove the swing *at random* time intervals, the overall effect will be almost no motion of the swing at all, even though you are certainly linked with it.

The idea of resonance has found its way into our everyday language also. We say that you pick up somebody's "vibes" when we want to express a great amount of overlap in feelings and thinking between you and another person.

Back to Resonant Vibrators

With the concept of resonance firmly in mind, we can understand the function of the resonant vibrator in instruments. The resonant vibrator of the instrument is the part placed in contact with the primary vibrator (which directly receives energy from outside the instrument). Its function is to accentuate certain frequencies that produce a strong and pleasing musical sound. For this reason the characteristic frequencies of the resonant vibrator are chosen to be those of musical sound, that is, twice, three times, and so forth, that of the fundamental frequency. In contrast, the characteristic frequencies of many primary vibrators in instruments are a motley group, and by no means form a musical sound. Take, for example, the reed of an oboe by itself, or the mouthpiece of a trumpet by itself, and activate it as if you were playing the instrument. The sound you will get is more a squeak than a musical sound. When you join the reed to the oboe or the mouthpiece to the trumpet, and then play the instrument, the whole spectrum of characteristic frequencies of these primary vibrators makes contact with the resonant vibrators of the instruments (which, in both cases, is the air column inside the tube of the instrument), but only those frequencies will be excited in the resonant vibrators that are their characteristic frequencies, and that are also represented among the characteristic frequencies of the primary vibrators. Thus the resulting sound will be a musical sound, since the characteristic frequencies of the air column are those of musical sounds, as we will see in later chapters.

There is also another function of this linking between the primary and resonant vibrators. As energy is fed from the primary to the resonant vibrators in the characteristic frequencies of the latter, the energy in the primary vibrator is redistributed to repopulate the frequencies energy has left to go into the resonant vibrators. As soon as this repopulation of frequencies occurs, more energy from them will pour into the resonant vibrator, hungry for such frequencies. Thus you can also think of the whole process as a rechanneling of the energy of the primary vibrator, originally distributed into many frequencies, into those particular frequencies that are characteristic of the resonant vibrator, and thus contribute to producing a musical sound. As a result the sound produced by the resonant vibrator not only will be more musical, but also will be louder. It is, of course, not a matter of getting more energy out of the resonant vibrator than the energy we transmitted to the primary vibrator. That would be a violation of the conservation of energy. It is only the rechanneling of the total energy imparted to the primary vibrator into frequencies of the resonant vibrator that contributes to the production of musical sound.

In different instruments we see different kinds of resonant vibrators. In instruments in the violin family, the resonant vibrating system consists of the top and the bottom plates of the instrument and the air volume inside the instrument. In all wind instruments, the resonant vibrator is the air column inside the instrument. In some brass instruments made of relatively thin metal, the walls of the instrument also contribute a little to the resonant vibrations, but in instruments such as the flute, oboe, and clarinet, the walls are much too thick to vibrate much.

In the piano the resonant vibrator is the sounding board, a large wooden sheet inside the piano body. In drums of various sorts, the resonant vibrator is the volume of air inside the drum. Some percussion instruments (such as the cymbal) do not have a special resonant vibrator, or, you might say, in the cymbal the plate functions both as a primary and a resonant vibrator. But other percussion instruments, such as the marimba, have special pipes in which the air columns act as resonant vibrators.

In the human voice, resonant vibrators play an unusually large and important role. They are the various cavities inside the head, that is, the throat, mouth, and nasal cavity. They take energy from the primary vibrators, which are the vocal cords, and select, reinforce, amplify, or dampen many frequencies to lend a highly individual character to the voice of each singer.

The Sound Effuser

Musical instruments are designed to produce a sound effect in the ears and minds of the listeners. Thus the vibrations, frequencies, and sound waves created in the instrument itself have to reach the listener as loudly as possible and without degrading their Fourier spectra so laboriously constructed by the player of the instrument. Each instrument, therefore, will have to make provisions for such a faithful and vigorous effusion of the sound created within.

One requirement for this is, of course, a medium that can vibrate and that fills the space between the instrument and the listener. Indeed, an orchestra playing in total vacuum would not be heard, not only because the players would suffocate, but also because the vibrations they produce would not carry from them to the audience. This is, however, seldom the problem. The crucial thing usually is to get the sound from inside the instrument to just outside it—the rest will be taken care of by the air (and by the acoustics of the auditorium, as discussed in Chapter 13).

In general, we need a hole in the material of the instrument through which the sound can escape, and we also want to provide as smooth a transition as possible between the inside and the outside, because the sharper the contrast between two media is, the more of the waves will reflect from the boundary. Different instruments solve this problem in different ways. The top of a piano can be opened for the sound to come out, and the violin and its relatives have two "f-holes" cut into the top plates of their bodies. Trumpets, trombones, tubas, and similar brass instruments have a flared end, thus making a tapered, gradual exit for the sound. Furthermore, all wind instruments use, as resonant vibrators, air columns in a tube *with one end open* (in addition to an end through which the instrument may be activated). If we paid attention only to creating a resonant vibrator, we could equally well use closed pipes. Guitars also have holes on their top plates to emit the sound.

Instrument Making: Science or art?

If you know something about the history of instrument making, you might be somewhat skeptical about our efforts here to understand the functioning of instruments on a scientific basis. You will remark that Antonio Stradivarius, during the 97 years of his productive life, or the Guarneris, or the Amatis, who, in the 16th, 17th, and 18th centuries, made the great violins we still use and have not been able to duplicate or surpass, probably knew little about the science of their times, which in any case was quite primitive compared with our scientific knowledge now. It is therefore obvious, you will say, that science is not needed to make great instruments, and hence, by extension, to understand how they work.

You are both right and wrong. To take this situation a bit further, we have to make a small excursion into a discussion of what science is, what technology is, and the relationship between the two.

Science and Technology

Science is a human activity the aim of which is to *understand and explain* the phenomena of the world around us. Once we have such an understanding, we can make *predictions* about the outcome of future phenomena, or present phenomena we have not studied yet. Thus the product of science is *knowledge, understanding,* and *predictive power.* Depending on the *motivation* for undertaking such an activity, we can distinguish between *basic* science, where the motivation is mainly to acquire such knowledge and understanding for their own esthetic and instrinsic values, and *applied* science, where the motivation is mainly the expectation that the knowledge and understanding can be utilized for some other human activity. Most scientific activity is both "basic" and "applied," in that the motivations behind it are a mixture of the two.

In contrast to science, technology is a human activity whose aim is to produce a way of doing or making something—a prototype, a process, a gadget. Thus the product of technology is something more tangible than the product of science.

The Interaction Between Science and Technology

Modern science was born around 400 years ago. All of the scientific knowledge accumulated up to that time was negligible compared with what we have accumulated in these past four centuries. The same can be said about technology, and so one might suspect that the two have interacted in this explosive growth over the past 400 years. To some extent this is so, but in other ways it is not.

When modern science began, it was occupied with the study of the natural phenomena we encounter directly in everyday life: falling objects, motion, forces, and the like. At that time, and even before, technology was also concerned, exclusively, with utilizing the properties of objects immediately in our environment. But science then was still young and not very powerful in its predictive power, and technology, in dealing with everyday objects, could make good progress by the age-old trial-and-error process, by just tinkering until something interesting and useful happened. So in this period technology was scarcely science based.

Soon, however, science had established our basic knowledge of everyday phenomena and moved on to the phenomena not so directly observable by our human senses, in everyday life, such as electricity and magnetism, later atomic phenomena, nuclear phenomena, elementary particles, distant astronomical objects, molecular biology, and many other areas. Parallel to this, technology also became satiated with purely mechanical devices built on the basis of tinkering with levers, wheels, and gears, and moved on to utilizing other, more "remote" natural phenomena. In doing so, however, it could no longer rely on trial and error, since we can tinker only with things we can handle directly and about which we have developed, through everyday contact over decades, an intuitive feeling of how they work. At this point technology had to involve, to an increasing extent, the newly generated scientific knowledge. For example, if you were presented with a box full of pieces of copper, glass, germanium, plastic, and so on, and asked to produce, by mere tinkering, a transistor radio, you would never be able to do so, since you would have to try, in the absence of scientific knowledge or intuition, billions of different combinations of the ingredients before you hit upon the one that would function like a transistor radio.

To be sure, even in the scientific age, technology is not completely free of tinkering or trial and error. Usually scientific knowledge is used to zero in on the approximate nature and structure of a new technological invention, and trial and error used for the final, detailed work.

All this is fine, you may say, but since instrument making has to do, from a scientific point of view, with vibrations, which is a mechanical phenomenon in principle established by science a long time ago, why did the famous violin makers of previous centuries not use that science for their craft?

In reply, one might point out that even in areas of science where the basic principles were established, in some instances a link with technology could not be made for a long time because the application of these basic principles to complex systems could not be carried out in practice, owing to the human inability to *calculate* so many components of a complex system simultaneously. In the past 50 years, however, electronic computers have been developed that can carry out an enormous number of simple calculations simultaneously and in an extremely short time. This has opened the door to using any scientific principle, even in complex systems where such a principle, together with many others, must be applied over and over and in parallel, to many components, in order to understand the system and predict its behavior.

Science and Technology in Instrument Making

Instrument making is a technological activity with a very tangible product, the musical instrument. Much of the historical manufacturing of musical instruments was done by trial and error and not by science-based technology. This is particularly true for instruments made of wood, since the technology of wood has not changed very significantly in the past 400 years. Thus violins and related instruments were made exquisitely a long time ago. As a result of scientific studies, we can now make violins on the basis of science-based technology that are quite good, though still not nearly as good as the best instruments made in the hayday of violin making 300 or so years ago. As computer-aided observations and calculations on the structure of violins become more extensive and sophisticated, the situation will improve, and there is no reason why we should not be able eventually to produce scientifically made Stradivaris.

The history of metal musical instruments is significantly different in this respect, since the (partly science-based) technology of metals has improved greatly in the past 400 years. Thus flutes, or even oboes or pianos (partly wooden), have greatly inproved over time, and today's metal instruments are in no way inferior, and in many instances are superior, to those made 200 years ago. In addition, even apart from the improvements in handling the material out of which the instruments are made, studies are also progressing on scientifically understanding the functioning and structure of these instruments, again promising steadily improved instruments for the future.

Furthermore, science-based technology has created entirely new instruments of the electronic type, for completely new potentialities in musical sound. Although many feel that the experimentation with these new sounds by present-day composers is still in a primitive state from an esthetic point of view, the future offers a conversion of these limitless opportunities into new realms of musical pleasure.

Therefore, in spite of Stradivari's creation of the greatest violins to date, the scientific understanding of musical instruments has a strong rationale and is indispensable for the future development of excellent musical instruments.

In our present discussions, of course, we will be able to touch only on the most elementary principles of these instruments. Such an elementary understanding, however, will prepare the reader to continue into the more detailed, sophisticated, and specific literature on instrument making.

Summary

All musical instruments consist of five main components: the energy source, the energy transmission to the instrument, the primary vibrator, the resonant vibrator, and the sound effuser. The efficiency of the conversion of energy to sound in all instruments is very low (1 percent or so). The primary vibrator converts the energy into vibrational energy, but neither its volume nor its Fourier spectrum is desirable. The conversion into the right kind of volume and Fourier spectrum is accomplished by the resonant vibrator. Every vibrating system has its own characteristic frequencies. When we feed energy into a vibrating system at its characteristic frequency, we create a resonance between it and us. In the resonant vibrator of an instrument, the energy of the primary vibrator is redistributed into the special characteristic frequencies of the resonant vibrator, which are such as to create a musical sound. The sound effuser carries the sound out of the instrument to the listener. To decide whether instrument making is science or technology, we specify that the product of science is knowledge, understanding, and predictive power, while the product of technology is a prototype, a process, a gadget. Several hundreds of years ago, science and technology were not so closely linked as today, because both dealt with phenomena directly accessible to human senses, and thus allowed tinkering by trial and error to produce new devices. This was also the case in the age when the great violins were made. In more modern times, technology has become much more science based. As a result we can make quite good instruments on the basis of scientific knowledge, and eventually science-based technology will produce more perfect instruments than was possible with the tinkering-based technology of previous centuries.

Chapter Nine

Wind Instruments

IN MOST BOOKS dealing with the physics of instruments, the order is: string instruments, wind instruments, percussion instruments. This order coincides with the order in which the orchestra members are listed on concert programs or located on the stage, but makes no sense at all from the point of view of understanding how these instruments work. From that point of view, wind instruments are the easiest to understand, and so we will start with them.

As we saw earlier, the energy source for such instruments is the muscles of the lungs and diaphram, which are responsible for the intake and outflow of air into and out of the lungs. (In a few instances, as with the electric organ, the energy comes from an electric motor that pumps air through the pipes.) The energy then is transmitted to the instrument by the flow of air, which carries a considerable amount of kinetic energy. Our detailed study of these instruments thus begins with the primary vibrator.

The Primary Vibrator

Three different kinds of primary vibrators are used in the various wind instruments: the vibrating reed, the vibrating lip, and the vibrating air in the so-called edge effect. In all three of these mechanisms, a crucial part is played by the same physical phenomenon—the so-called Bernouilli effect.

The Bernouilli Effect

The name of this phenomenon is awe-inspiring, but its essence is very simple, as one can see easily with no more equipment than a sheet of paper and the lungs.

Take that piece of paper, and hold your head so that you blow the air out of your mouth in a horizontal direction. Hold the paper just *below* your lower lip, against your face, so that the part of the paper nearest to your face is horizontal (see Figure 9.1). The rest of the paper, of course, will curve down, and the end of the sheet will probably be almost vertical, because its weight pulls it down at the unsupported end. Now take a deep breath and blow air out of your mouth uniformly and quite strongly for a few seconds. As soon as you start doing this, you will see the hanging end of the sheet of paper rise to the point that the whole sheet is in a horizontal position. The paper will stay in that position as long as the strong airstream continues to flow from your mouth. When you stop, the paper drops into the original, hanging position.

It is not difficult to conclude that when the airstream passed over it, the paper popped up from its hanging position and apparently moved against the gravitational force pulling it down because another, greater force was pulling it up into that horizontal position. That force was a lateral suction due to the fast moving airstream. This phenomenon, called the Bernouilli effect, is named for the Swiss scientist Daniel Bernouilli, who presumably first noticed it in the 1730s.

The Bernouilli Effect In Reed Mouthpieces

Many wind instruments, such as the oboe, clarinet, and bassoon, have, as primary vibrators, mouthpieces made of one or two pieces of flexible reeds. In the double-reed mouthpieces, the two pieces of reed are close together with only a small opening between

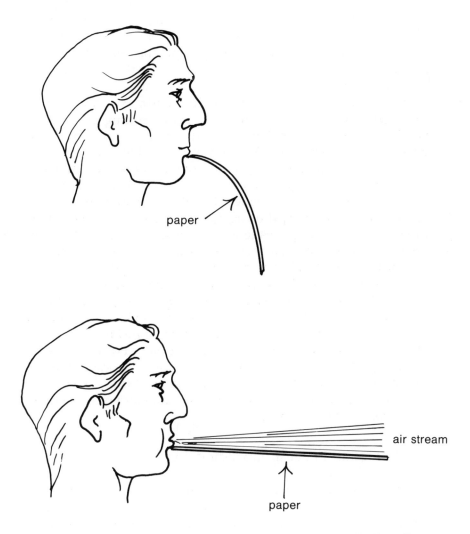

Figure 9.1 Do-it-yourself experiment to illustrate the Bernouilli effect. If you blow *over* a paper fitted to your lower lip, the paper will rise as long as the air-stream passes over it.

them, and they both can bend so as to make the opening larger or close the opening altogether. In a single-reed instrument, the reed is close to the wall of the instrument, the latter being rigid, but the one piece of reed can flex to make the opening larger or smaller. From our point of view, there will be no essential difference between single-reed and double-reed instruments.

In either case, when we blow into the mouthpiece, we produce a fast airstream that passes the reed(s), thus setting up the Bernouilli suction in a lateral direction; that is, perpendicular to the direction of the airstream. As a result the flexible reed(s) move(s) and close(s) the opening, and the airstream stops—and with it the Bernouilli suction. Hence there no longer will be a lateral force acting on the flexible reed(s), which, therefore, will jump back to its (their) original position, thus again creating an opening through which the airflow can resume. This process repeats itself extremely rapidly, hundreds or even thousands of times per second, so that the reed(s) will be in constant vibration with audible frequencies, as long as the player blows into the mouthpiece.

The vibration of the reed(s) will be only partly periodic, since the details of the airflow and the way the Bernouilli suction affects the various parts of the reed(s) can vary from instant to instant. Thus the sound emitted by the mouthpiece alone is a semimusical squeek. But this will be of no concern to us since we know that the actual musical sound the instrument produces for our listening is determined in cooperation with the resonant vibrator, and not by the primary vibrator alone.

The Bernouilli Effect in Lip Vibrations

Other wind instruments, such as the trombone, French horn, trumpet, and tuba, do not have vibrating reeds in their mouthpieces. Indeed, there is apparently nothing elastic or flexible at all in their mouthpieces: they are made of solid metal. In such instruments the primary vibrator is the lips of the player.

The mechanism is very similar to that in the reeds. The player forms a very small opening with the lips through which the airsteam flows. The flow sets up a Bernouilli suction that sucks in the the soft lips to close the opening, thus stopping the airflow. This, in turn, stops the suction, the lips spring back into their original position, and the airflow resumes again.

The Bernouilli Effect in the Edge Effect

A few wind instruments, such as the flute, piccolo, recorder, and certain types of pipe organs, use the Bernouilli suction, together with other effects, in yet another way in their primary vibrators. Consider an obstacle with an edge in the path of an airstream (Figure 9.2). Let us assume that the airflow is mainly below the edge. This airflow, through the Bernouilli effect, sucks some more air from the volume just above the edge into the space below the edge. The excess air there creates whirls of air, which tend to obstruct the airflow, and some of the excesss air also flows around (see figure) and tries to go to the space above the edge where there is a shortage of air (since the air was sucked in from there). This circulating air tends to push the airstream above the edge, so that in the next instant the airstream will be above the edge, and then the whole process repeats itself, and the airstream oscillates very rapidly from above to below the edge and back again. The exact mechanism of this oscillation is yet to be understood, though we know that these factors all play some role in it. The result, from our point of view, is a rapid oscillation of the airstream. In this case, therefore, the primary oscillator is not a mechanical part of the instrument but the airstream itself.

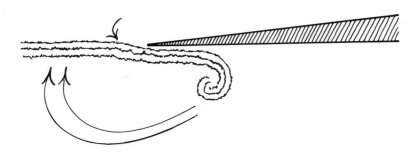

Figure 9.2 The edge effect. The airstream blown at an edge first passes, say, below the edge, but then the combination of the Bernouilli effect and eddies flip it over to above the edge. Then the same effects flip it again to below the edge, etc., thus creating a vibration of the airstream.

Summary of the Various Types of Primary Vibrators

We have seen that in the wind instruments three kinds of primary vibrators can be found. All of them depend on the Bernouilli effect, though the edge effect also involves other physical phenomena. The Bernouilli effect is the lateral suction set up by a fast-moving airstream. In reed instruments this lateral suction alternately bends and releases one or two reeds, the vibrations of which form the primary vibrations. In brass instruments the lateral suction affects the player's lips and makes them vibrate. Finally, in the group of instruments that utilize the edge effect, the primary vibrator is the airstream itself that alternately passes above and below an obstructing edge placed in the path of the airflow.

The Resonant Vibrator

In all wind instruments, the resonant vibrator is the air column inside the tube of the instrument. In almost all cases, the diameter of these tubes is very small compared with the length of the tube (less than 5–10 percent), and hence these tubes, to a good approximation, can be regarded as one dimensional, which makes their discussion very simple. Small corrections to these results can be calculated, taking into account the small but not completely negligible other two dimensions of the tube as well, but we will not worry about those corrections here.

The discussion of the various vibrations that can occur in such a one-dimensional tube is extremely simple. At a closed end of the tube, we must always have a node of the standing wave; that is, we must have zero amplitude for the vibration there at any time. This is so since the wall closing the tube is rigid and cannot vibrate and hence the air immediately in contact with it cannot either. At an open end, in contrast, we must always have a maximum amplitude, since there nothing restricts the vibration. These two rules then automatically determine the various vibrations we can have in a tube. They are shown in Figure 9.3. Note that there are three different cases: both ends closed, one end closed and one open, and both ends open.

Tube with both ends closed.

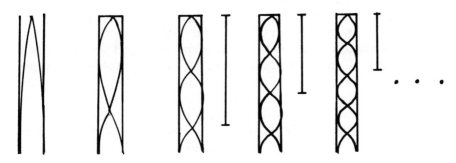

Tube with one end open and one closed.

Tube with both ends open.

Figure 9.3 The standing waves that can exist in air-filled tubes. The bars show the wavelengths of the various waves. A closed end must always have a node, while an open end must always have a maximal vibration.

We see that the kind of wavelengths that can occur in such a tube is determined only by the length of the tube. In looking at Figure 9.3, we see immediately that for the two cases when the ends are either both open or both closed, the wavelengths of the waves can be twice the length of the tube, or half of that (i.e., half of twice the length of the tube, which happens to be just the length of the tube), or one-third of twice the length of the tube, or one-quarter times twice the length of the tube, and so on. The wavelengths are indicated in the figure by brackets at the right of the tube. A few of the longest wavelengths are not shown so that the reader can work these out as an exercise. We see, therefore, that in the case of tubes with both ends open or both ends closed, the basic distance scale unit is *twice* the length of the tube, and the possible wavelengths can be that unit, or one-half of it, or one-third of it, or one-fourth of it, and so forth.

The situation for tubes with one end open and one closed is a bit different. From the figure you can see that for the longest wave possible, the tube contains one-fourth of the whole wave; for the next, three-fourths; for the next, 1¼; for the one after that, 1¾; then 2¼, and so on. Since we see quarters everywhere, we can say that the basic scale unit here is *four times* the length of the tube, but then the possible wavelengths are that unit itself, one-third times it, one-fifth times it, one-seventh times it, and so on. In other words, we can have only *odd* numbers in the denominators.

Do *not* try to memorize these two rules. They can be deduced simply from the figure. These, in turn, can be constructed simply from our two rules governing what happens at the open and closed ends of pipes, so that whenever you need these results, you can reconstruct them in a minute. If you try to memorize them, however, you may recall them incorrectly, without realizing that you have done so.

The foregoing tells what wavelengths the standing waves, and hence the vibrations of the air column, can have in the resonant vibrators of wind instruments. To convert this into pitch and Fourier spectrum, however, we need frequencies, not wavelengths. It is at this point that the wind instruments show great simplicity, because what is vibrating in their resonant vibrators is an air column, and since we know the speed of sound in air, we can immediately convert these wavelengths into frequencies. Consider an example. We have a tube, three feet long, with one end open and one closed. What are the frequencies of the characteristic vibrations of the air column in this tube? From our figures we see that the various characteristic standing waves in such a tube have wavelengths of 12 feet, 4 feet, 2.4 feet, 1.71 feet and so on. Since the speed of sound is 1000 ft/s, the corresponding frequencies are 83⅓ per second, 250 per second, 416⅔ per second, 583⅓ per second, 750 per second and so on.

Remember, however, that the speed of sound in air depends somewhat on the temperature of the air. The speed is greater as the temperature rises. As we have seen, in wind instruments the *wavelength* of the sound the instrument makes is determined by the geometrical dimensions of the instrument. The corresponding *frequency,* however, which determines the pitch we hear, will depend on the speed of sound also, and hence will change when the temperature changes. As the temperature rises, the speed of sound increases, and hence, for a given wavelength, the frequency will also increase, and the pitch becomes higher. This is the reason why wind instruments tend to get "out of tune" in a concert hall, even if at the beginning, when the temperature of the air was still lower, they were "in tune" (as compared with instruments such as the violin, which are not so affected, or at least not in the same way).

The Practical Problems with Resonant Vibrators

The problem with having a simple, single tube as a resonant vibrator is evident: the variety of notes such a tube can produce is extremely limited. In general, in fact, we will have only one single note and its overtones, with the single note determined by the length of the tube. If we are lucky, we may be able to excite the primary vibrator in such a way that it produces vibrations in only *some* of the possible characteristic frequencies of the resonant vibrator, so that we can produce more than one note. Indeed, this can be done in most reed instruments by "overblowing," that is, by blowing harder than usual, in which case one usually excites an overtone that is one octave higher than the normal tone we get by blowing only moderately. In brass instruments it is possible for a very skillful player to change the lip position in various ways to excite only certain characteristic frequencies of the resonant vi-

Flute (Courtesy
the Selmer Company,
Elkhart, Indiana)

Oboe (Courtesy
the Selmer Company,
Elkhart, Indiana)

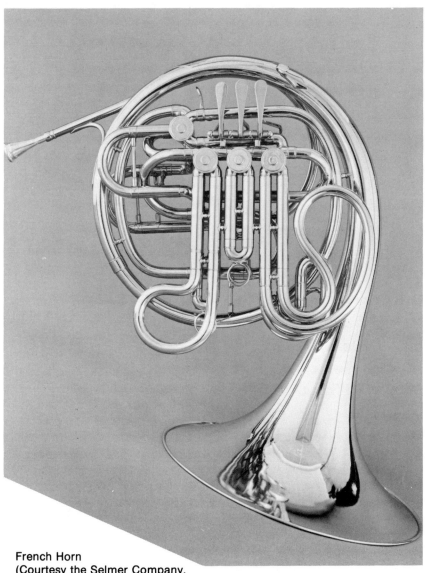

French Horn
(Courtesy the Selmer Company,
Elkhart, Indiana)

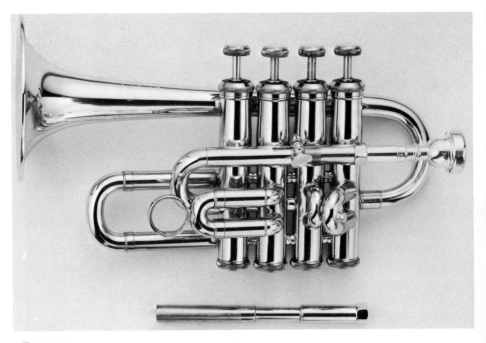

Trumpet
(Courtesy the Selmer Company,
Elkhart, Indiana)

Trombone
(Courtesy the Selmer Company,
Elkhart, Indiana)

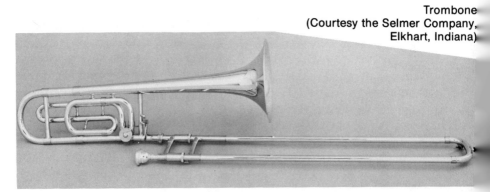

brator—which is the method used to play valveless trumpets and other valveless brass instruments. But even if we employ all these "tricks," the set of notes playable on a tube of fixed length remains very limited, and certainly does not form a scale. On a tube with one end open, for example, we can produce the fundamental note, then the first harmonic with three times the frequency (which is higher than the fundamental by one octave plus five notes on the major scale), then the second harmonic, which has five times the frequency of the fundamental (which means higher than the fundamental by two octaves plus three notes on the major scale), and so forth. Such intermittently located notes can hardly serve as a basis for interesting music. The situation with tubes with two ends open is a bit more favorable, but still inadequate.

Since the frequency produced depends only on a single parameter, (the length of the tube), there is only one remedy: we must have access to a variety of tube lengths in order to produce a sufficiently large number of notes.

The Solutions

There are *four different solutions* used to bring about this remedy, namely to offer access to a variety of tube lengths. They are: (1) many tubes; (2) interlocking pipes; (3) temporary extensions with switches; and (4) holes on the sides of the tubes.

Solution 1. Include Many Tubes of Differing Lengths

This is clearly the "brute force" solution, the option of one who is unimaginative but rich. Such a solution results in an instrument that is large and bulky, and requires much raw material and labor to manufacture. The solution is used only in one instrument—the organ. But there it serves other purposes also, in addition to offering a variety of pitches. In the organ there are hundreds, or even thousands, of different pipes. They differ in length, but also in other respects, so that the Fourier spectra of the sound even for two pipes of equal length are different. In some certain overtones will be prominent; in other pipes different overtones will have large Fourier amplitudes. As a result the timbre, the tone quality of the two pipes, will differ. Thus the organ offers a huge variety not only of pitches, but also of timbres, which compensates for its great bulk and very high cost. The organ is the most expensive instrument of all; its cost can exceed even that of a rare Stradivarius violin by a factor of five.

Solution 2. Vary the Length by Interlocking Pipes

The principle of this is shown in Figure 9.4. This geometrical arrangement is actually used in the trombone, which is the only present-day instrument utilizing this kind of a solution. Two pipes fit into each other and can slide within each other, thus allowing us to vary the combined length of the tubing. There are two problems with this solution. One is that the tubes can change their lengths continuously, and correspondingly the pitch of the note can also change continuously. This is fine if we want to produce "sliding" notes, but these are seldom required. Most of our music, as mentioned earlier, uses fixed-frequency notes arranged in a scale, and so the trombone player has to be skillful to learn which particular extensions of length correspond to the actual notes on the scale.

The second problem with this solution is that, in practice, the range of changes in length one can achieve is very limited. After all the player's arm is just so long, and whether the arm is folded or extended determines the minimum and maximum lengths one can achieve. On a trombone the range of pitches possible with *sliding alone* is only six semitones, or about half an octave. The trombone actually has a larger range of notes than that, but the extra range is achieved with overblowing and changes in lip position, as mentioned earlier.

The changing of length by sliding is also a bigger, more awkward motion of the human body than the motion of the fingers, which is used, on other wind instruments. As a result the trombone cannot play a sequence of notes as fast and as flexibly as some other wind instruments.

Solution 3. A Switchyard with Temporary Extensions of Length

Consider Figure 9.5. Here we have a tube, which somewhere has a bifurcation into two alternative branches that soon join

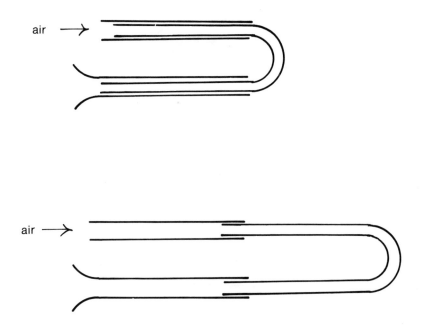

Figure 9.4 Regulation of the tube length for a trombone using two separate tubes that fit into each other.

again, and these two branches have unequal lengths of tubing in them. We also have a valve at the junctions so that we can direct the airflow into either one branch or the other. It is clear that the tube with one branch inserted in it will have a different overall length, and hence different characteristic frequencies, than the same tube with the other branch inserted into it.

We can make things even more variable by inserting several such bifurcations (and thus several valves) into the same tube, and thus offer a greater variety of tube lengths through the various *combinations* of settings of the valves.

This solution is used in most present-day brass instruments, such as the valved trumpet, the tuba, and the French horn. In most cases this solution is also used in *conjunction with* influencing the pitch through the strength of the airstream and the lip position.

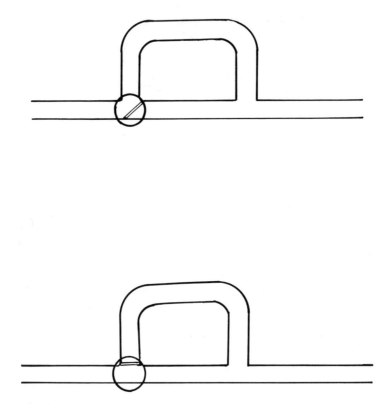

Figure 9.5 Regulation of the tube length with the help of a "bypass" valve. In the upper position, the air is deflected through the bypass; in the lower position, the air goes through the shorter straight section.

Solution 4. Place Holes on the Side of the Tube

We can also vary the length of the tube by drilling a set of holes on its side, along its length, and cover or uncover these holes, depending on what pitches we want to produce. The covering and uncovering can be accomplished by small and easily manipulated covers attached to levers that can be activated with the fingers, or by the fingers themselves. This solution is used for all reed instruments, such as the oboe, clarinet, bassoon, and saxophone, as well as on those instruments that utilize the edge effect, such as the flute, piccolo, or recorder. It has the advantage of easy handling, and since players have at least eight fingers at their disposal (two may be used to hold the instrument), the variety possible is great.

Since an octave corresponds to a factor of two in length, in this solution there is a limit as to what fingers alone can do, since they cannot reach very large relative distances. Thus in these instruments also, fingering is used in conjuction with overblowing, so that the finger holes themselves need to take care of only one octave.

The positions of the holes on the sides of the tubes are not exactly those on the very simple model we have been using. A tube with a hole in its side does not become exactly like a tube that ends at that hole, since the airstream can, in part, continue down the tube instead of exiting through the hole. These finer details, therefore, have to be treated with much more sophisticated mathematics combined with trial-and-error-type experimentation.

Another reason why the hole positions are not exactly where our simple model would place them is that, in actuality, the tube of musical instruments (called the bore on the woodwinds) is not quite cylindrical; its diameter increases slightly away from the primary vibrator. In other words, the instrument is actually slightly conical. This also introduces small deviations from the way things could be for the idealized instrument we have considered, namely, a cylindrical tube or tubes with negligible diameters.

The Effuser

The sound comes out of a wind instrument through the hole at the end of the instrument opposite to where the primary vibrator is located. If the tube of the instrument were unchanged in shape to its very end and the transition to the air outside would thus be abrupt, at the end of the tube a sizeable fraction of the sound wave would reflect back into the tube instead of going out of it, and hence the volume of the sound going to the listener would be relatively small. If, on the other hand, the transition to the outside air is made gradual by attaching a flare to the end of the tube, the reflection is less and hence the sound going to the outside increases in volume. This process is quite analogous to what in electronics is called impedance matching.

Since the sound leaves the instrument primarily along the direction of the axis of the tube, even if it spreads around in other directions afterwards, the instrument remains highly directional in that the volume of sound along the continuation of the axis of the instrument is larger than in other directions. This is recognized, for example, in the placing of musicians in the orchestra. Wind players are usually located in the back of the orchestra, facing the audience, so that the axis of their instruments points straight into the audience. In some cases this is not possible: The way the flute is played (from the "side"), it is not possible to point the axis to the audience and to have the player face the conductor at the same time. But the flute, usually playing at high frequencies, can take care of itself even with some handicap. There is more of a problem with French horns, so one may need more of them to balance the orchestra.

Summary

The primary vibrator in wind instruments functions on the basis of the Bernouilli effect, although in such instruments as the flute and recorder other effects also are in evidence. The Bernouilli effect consists of a fast-moving airstream that creates a lateral suction. In primary vibrators that use one or two reeds, the vibration of the reed is created by the airflow inducing the Bernouilli effect to act on the reed, which then closes the passage for the airflow. This stops the Bernouilli effect, thus opening the passage again, and so on. In other wind instruments, there is no reed but the Bernouilli effect acts, in the same way, on the lips of the player. In the case of the flute and its relatives, the primary vibrator is an airstream that is oscillated by the edge effect, due to the obstruction of an edge in its path.

The resonant vibrator in all wind instruments is an air column inside a tube. The possible wavelengths of standing waves, and hence the frequencies of the possible vibrations, can be very easily constructed in such a tube by remembering that a closed end of the tube must always have zero amplitude of vibration whereas an open end must have a maximum amplitude of vibration. The result: For a tube closed at both ends or open at both ends, the wavelength is equal to twice the length of the tube, or one-half of that, or one-third of that, and so on. For a tube open at one end and closed at the other, the wavelength is equal to four times the length of the tube, or one-third that, or one-fifth that, and so on, using only the odd numbers in this sequence.

This spectrum of frequencies is quite sparse. To increase the number of notes available on such instruments, four different solutions can be used: (1) to acquire access to many pipes of differing length; (2) to lengthen or shorten the tube by two interlocking tubes sliding within each other; (3) to change the length by a series of bifurcations into branches of unequal length, activated through valves; (4) to place holes on the side of the tube and cover and uncover them in the course of music taking.

The effuser in wind instruments is the flare at the end of the tube. Wind instruments are highly directional.

Chapter Ten
String Instruments

I<small>N DISCUSSING WIND</small> instruments in the previous chapter, we spent much time on the physics of the resonant vibrator and little on the primary vibrator system. In this chapter the opposite will be true—most of it will be devoted to the primary vibrator system. It is not a matter of which is more important. We know by now that the two vibrators must work together on an equal footing in order to create great sound. The reason is that in the case of string instruments, it is the primary vibrator that can be subjected to a relatively simple physical analysis, while a similar study of the resonant vibrator is much more complicated, as we will see. In wind instruments it was the other way around. Although we discussed the overall principle of the primary vibrators in the form of the Bernouilli effect, the detailed study of the way reeds or lips vibrate, or the way the airflow oscillates in the edge effect, is much too complicated to be appropriate for such a survey. On the other hand, the resonant vibrator in the wind instruments is quite simple and can be treated quantitatively even in an introductory text.

The first two components of such instruments have already been mentioned for string instruments. The energy is produced by the arm muscles of the player for the violin family, and/or the finger muscles for almost all string instruments. The transmission from this energy source to the instrument is arranged differently in various string instruments. For the violin family, the bow has hairs attached to it that are treated with a substance that makes them rather sticky. When the bow is drawn across the strings, it catches a string, and pulls it out of its resting position. Then, when the resistance of the string becomes too large, the string breaks away from the bow, and jumps back toward its rest position, only to be caught again by the sticky bow, released again, and so on. In string instruments such as the guitar, harp, mandolin, and lyre, the transfer is directly through plucking with the fingers. In the case of the piano, the transfer is through the fingers depressing a key that activates a composite of levers that eventually moves a hammer that hits a string. In some of the predecessors of the piano—the harpsichord, for example—the key activates something that plucks the string inside the instrument.

The Primary Vibrator

The primary vibrator in a string instrument is the string (almost always several of them), fixed at two ends but otherwise free to vibrate.

The kind of standing waves and their *wavelengths* that can be generated on such a string can be pictured very easily. As we will see, the analysis of the corresponding *frequencies* is more involved. But let us first discuss wavelengths.

Such a discussion is similar to, but even simpler than, that of the standing-wave patterns in the resonant vibrator of the wind instruments. The principle is the same, but the situation is simpler; with the string *both* ends are always fixed, which corresponds in the wind instruments to one of the three cases discussed (a tube with both ends closed). Thus the kind of patterns we can get on a string are similar to the first row alone of Figure 9.3. In the case of the string, these are as shown in Figure 10.1. The rule giving the wavelengths of the possible standing waves is also similar. These wavelengths can be twice the length of the string, half of twice the length, one-third times twice the length, and so on. That is all there is to the discussion of the *wavelengths* of the possible standing waves, at least in the somewhat idealized case where the string is thin and completely flexible.

etc.

Figure 10.1 The waves that can be generated on a string with both ends clamped. The ends must be nodes.

For the resonant vibrators of *wind* instruments, once the wavelengths were determined, finding the frequencies was easy, because the vibrations were in air, in which we know the velocity of sound, and hence there was a unique and simple relationship between wavelengths and frequencies. This is not true for strings. Let us consider our everyday intuition regarding the rapidity of such vibrations in a string. We can easily understand that even if we compare various strings of the same length, the rapidity of the vibrations will not always be the same, but will depend on at least two other factors.

The first of these is the degree of tension the string is under. When we pluck the string and thus remove it from its usual position, we make it bend and also stretch a bit. If the string is made of an elastic material (as it always is on musical instruments), this bending and stretching will induce an elastic stress. If the string were stretched more tightly to begin with, the stress under such deformations would also be greater. When this stress is greater, the string, when released in a deformed position, will try to get back to its usual position faster than if the stress were less. Returning faster (and then overshooting and starting an oscillation around the rest position) means that the frequency of the vibration will be larger.

The second factor that will influence the rapidity of the vibrations is the weight of the wire (say, per unit length) out of which the string is made. It is plausible that a wire that weighs more, will respond more sluggishly to a given stress than a string made of lighter material that is subjected to the same stress, and hence to the same forces trying to bring it back to its rest position. Thus the frequency of the vibration of a heavy string will be lower than the frequency of a light string of the same length and under the same stress.

Since the length of the string uniquely determines the *wavelengths* of the standing waves we can have in that string, but the *frequencies* depend, not only on the length of the string, but also on two other properties (the tension on the string, and the heaviness of the material per unit length out of which the string is made), the speed of propagation of waves in the string's material must also depend on these two additional factors. In any case we see that for the vibrations of strings, the relationship between wavelength and frequency is much more complicated than for vibrations in air. To summarize, the wavelengths on the string are uniquely determined by the length alone, but the corresponding frequencies also depend on the tension in the string and the weight of the string material. If we keep these two other factors fixed, then the relationship between wavelengths and frequencies becomes simple in that a wave with twice the wavelength will have half the frequency, and so on.

The discussion in the last two paragraphs can be rephrased in more scientific terms, using the concepts of force, acceleration, and speed—which we introduced in our general study of motion. If we displace the elastic string from its equilibrium position, there is a force trying to restore the string into its original equilibrium position. The tighter the string, the greater the restoring force. For a given mass, force is proportional to acceleration. Thus a greater tension in the string produces greater restoring force which in turn produces a larger acceleration, so that the string picks up greater speed in a given length of time and hence will vibrate faster. For a given amount of force, however, the larger mass the string has (per unit length), the smaller acceleration it will have, and hence its vibration will occur with a smaller frequency.

Whereas this more complex situation is perhaps a pity from the physicist's point of view, it is an advantage for the musician, since the increased number of parameters adds flexibility and simplicity to the construction of instruments. For example, the modern violin has four strings of *equal length*, which makes the construction of the violin easier. Yet the purpose of the four strings is to create a larger range of frequencies. If we had four air pipes instead of four strings, the four pipes of equal length would, by necessity, have the same characteristic frequencies, thus making three of them superfluous. With the four strings, however, this is not so, since we can place them under different tensions and make them different in thickness (and hence in weight per unit length), and in that case their characteristic frequencies will also be different, even though the strings are of equal length.

Frequency Range for String Instruments

As Influenced by Tension and Weight

We have just seen that the frequencies of sound produced on strings are influenced not only by the length of the strings, but also by the tension under which strings are stretched and by the weight per unit length of the material out of which the strings are made. It should be noted, however, that these two ways of regulating the frequencies, while very helpful when building the instrument or when preparing it for a musical performance, are practically useless for a *quick* change of frequencies while a musical piece is being played. To change the tension on the string in a particular way so as to produce a particular frequency takes at least 10–15 seconds even on an instrument that is suited for this purpose, such as the violin, and much longer on an instrument such as the piano. As for changing the weight of the string per unit length, whether by changing the material or by changing the thickness of the string, this takes minutes even in the most favorable case. Thus, as we will see, to change frequencies during a musical performance, we are reduced to changing the length of the strings.

Yet the option of influencing frequency when building the instrument or when preparing for a performance (i.e., "tuning" it) through using different tensions and different weights per unit length is extremely helpful. We already mentioned the example of the four strings of the violin, all of equal length but giving very different frequencies. A second example is that of the piano, in which each frequency is produced by a different string. The piano has an unusually large range of frequencies; on a concert grand, this can be more than seven octaves. Let us assume that we wish to achieve this entirely through the lengths of the strings. In that case the ratio of lengths of the strings producing the highest and lowest frequencies would have to be $2^7 = 128$. Let us make the string of the highest frequency four inches long. This is really the best we can do. Strings shorter than that, and with the usual thickness, have a very distorted overtone spectrum and sound bad. But even with this string of minimal length, the length of the string producing the lowest frequency would have to be 128 times longer, or be about 43 feet long. An instrument of that size would be extremely difficult to play.

As a third example of what would happen if the frequency of string instruments were determined only by the length of the strings, consider the instruments of the violin family— the violin, viola, cello, and double bass. Within the family we want to represent a large range of frequencies in order to give more variety to the musical pieces that can be performed by a string quartet or an orchestra. Indeed, the basic frequency of the highest string of the violin is four octaves higher than the basic frequency of the lowest string of the double bass. (Here "basic" means the frequency of the "empty" string, i.e., without any fingers being placed on the string.) The length of the violin strings (from the bridge to the upper end of the neck) is about a foot and a half. This is determined mainly by the size of the human arm and the mechanics of playing the violin. If we wanted to make the double bass with strings of the same thickness, material, and tension as those on the violin, the lowest string on the double bass would have to be $2^4 = 16$ times longer than that of the violin, or about 24 feet long, making the instrument completely unsuitable for practical use.

We see, therefore, that in building string instruments and preparing them for performance, the freedom of influencing frequencies not only by length, but also by tension and the weight of the strings, is indispensable. This makes string instruments somewhat more complicated, but also provides a greater number of options in structure and mechanics than we saw in the case of wind instruments where the frequency (in that case, of the resonant vibrator) was completely determined by the one parameter of length.

Frequency Range for String Instruments: Influencing by Length

In order to have a larger variety of frequencies, and also to have a rapid method of changing frequencies, all string instruments also involve the dependence of frequency on the length of the string. There are basically two ways to achieve a variation in length.

Solution 1. Having Many Strings with Differing Lengths

In a way this is the same kind of a "brute force" method as the first solution we saw in the case of wind instruments, where we utilized many tubes of differing lengths. But in the case of string instruments, this is a necessity if one wants a large range of frequencies on the same instrument, since, as we will soon see, the second method can change the length of a given string only by a factor of four or so, at the most, thus covering only two octaves, which is insufficient for many purposes. In any case, however, if we want to have a string instrument with a very large range of frequencies (such as the seven octaves of the piano), we must have a large number of different strings. This is the option used for the

Cello
(Courtesy the Selmer Company,
Elkhart, Indiana)

Violin
(Courtesy the Selmer Company,
Elkhart, Indiana)

piano, which has a separate string (or in many cases two or three strings) for each frequency. For reasons mentioned in the previous section, these strings are *not* under the same tension, *not* made of the same material, and are *not* of the same thickness. To allow the strings for the low frequencies to be made shorter, the weight per unit length of those strings is made much larger than for the high-frequency strings, and the tension is lowered somewhat. The weight per unit length is *not* increased by making the string out of a solid wire of larger diameter, because such very thick, and hence not fully elastic, wires would not have a very musical series of over-tones. The wavelengths of the possible standing waves are still as discussed earlier, but the relationship of the frequency of the vibration to the wavelength of the standing wave would for such a thick wire be different for each overtone, even when the tension and the thickness are the same. To avoid this the wire used for the string is only of moderate thickness, but a thin wire is wound around the string, to add to the weight per unit length without degrading the elastic properties of the string.

Another instrument that uses this method is the harp, which also has a large number of strings of varying lengths and thickness.

There is an interesting technological sidelight to such instruments. In the piano or harp, each string must be given a fair amount of tension in order to produce a musical sound; a floppy string does not vibrate, or vibrates only in a very irregular and un-uniform way. These tensions on the individual strings—which, on the harp or piano, are all attached to the same frame—add up when it comes to calculating the *total* tension on the frame itself. It is not difficult to conclude that with each string under a tension of several 10s of pounds, and with 100 or more strings, the total tension on the frame is *several tons*. The frame, therefore, must be made of strong metal to withstand such a force without deforming. This is one (though not the only) reason why the modern piano was born only about 200 years ago. It was necessary to have the technology to make such strong metal frames in an efficient way before such an instrument could be created.

The piano did have predecessors, such as the harpsichord, but this latter instrument has fewer strings, and each is under a smaller tension. As a result the total force on the harpsichord frame is much less than that on the piano frame, and so wood can be used in its construction.

Solution 2. Pressing a Finger on the String

If the string is stretched out close to a solid surface (the finger-board on the violin), then pressing the string in a particular spot against this surface makes it in effect shorter, since the part of the string beyond this finger position will not be participating in the vibrations at all. Needless to say, this method can only shorten the string (and not make it longer), and hence it can only raise the frequency (and not lower it).

This method is used in most present-day string instruments such as in the members of the violin family, the guitar and its relatives, and the Japanese koto. The great advantage is that the fingers of an experienced player can move very fast, and so the changing of the frequency can take place very rapidly—perhaps four or more times a second. The fingers have to move smaller distances on a smaller instrument, and therefore the violin has, in this respect, much greater flexibility than, say, the double bass.

In addition, all instruments that use fingers to change frequencies have several strings which helps to extend the frequency range. The number of strings, however, need not be large: there are four on modern violin-type instruments, six on the guitar, five on some older versions of the violin family, and so on.

Location of Energy Transfer

The energy transfer to the string can be made, in principle, anywhere along the string. For example, when one plucks a guitar string, one can do so at various spots along the string; similarly, one can bow the violin string anywhere between the bridge and the neck. Does it make a difference where the energy is transferred? It does as we will now see.

Depending on where the energy transfer is made, a different admixture of the (same) set of characteristic frequencies is excited. In other words, the Fourier spectrum, although involving the same frequencies, will have a different set of strengths for these frequencies, depending on where one transfers the energy. This is easy to see.

Figure 10.2A shows the first 10 harmonics on a string of a given length. In Figure 10.2B, the relative strengths of these 10 characteristic frequencies are shown for five different plucking positions on the string. The small figures at the top of Figure 10.2B show the plucking position, which divides the string in the ratios of 1/2:1/2, 1/3:2/3, 1/4:3/4, 1/5:4/5, and 1/10:9/10, respectively, in the five cases we are considering. The actual bowing position used on the instruments of the modern violin family corresponds approximately to the last of these cases, namely, 1/10: 9/10.

Below the small figures showing the plucking positions are the relative strengths of the nine harmonics corresponding to 2, 3 . . . , 10 on Figure 10.2A. These relative strengths are all given with respect to the strength of the first harmonic, which is taken to be 1.0, and is not shown. In actuality some of the harmonics have to be admixed with a negative sign, but this is not illustrated as it is not important for our present discussion.

There are two interesting features one can easily discern on the figures.

First, in each case the contributions to the admixture of those harmonics that have a node at the plucking point are always zero. In other words, these harmonics are missing from the Fourier spectrum altogether. You can see this by looking, in Figure 10.2A, at those harmonics that are missing in the various Fourier spectra. For example, for the 1/2:1/2 case, all even-numbered harmonics are missing. For the 1/3:1/2 case, those bearing the numbers 3, 6, 9, and so forth, are missing.

The second interesting feature is that the closer we pluck to one end of the string, the more important a role the higher harmonics play as compared with the first one. This is evidenced not only by the growing size of the second harmonic as you go from the 1/3:2/3 case to the 1/10:9/10 case, but also by the similar growth (on the average) of the higher harmonics. This is easy to understand. In the 1/10:9/10 case, the position of the string changes rapidly from the end (where it is at zero displacement from the rest position) to the plucking position only one-tenth of the length away (where it is a maximum displacement). Such a rapid change could not be described with a mixture of waves each of which changes only slowly, as the lowest harmonics do. Hence to describe such a rapid change, we must have a considerable admixture of the rapidly changing higher harmonics.

The resulting sound demonstrates the practical implications of this difference. If we bowed halfway along the string, the sound,

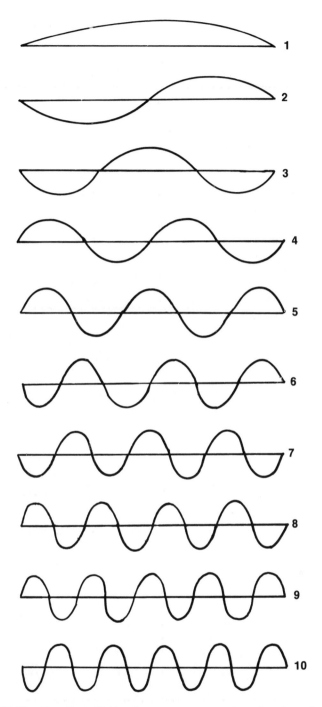

Figure 10.2A The first 10 Fourier components of a vibration of a string clamped at both ends.

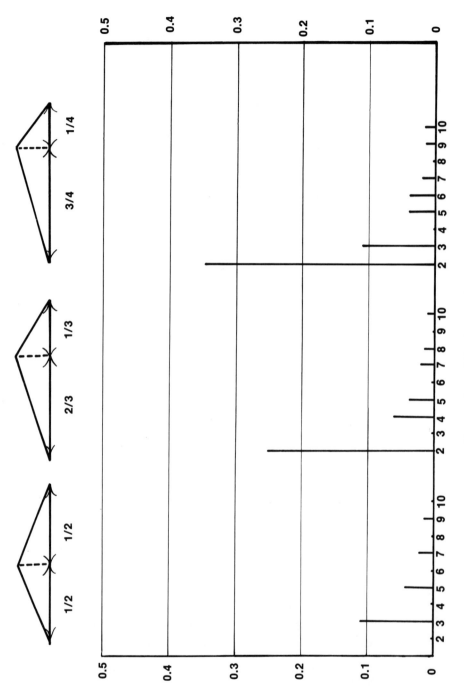

the number of the harmonic

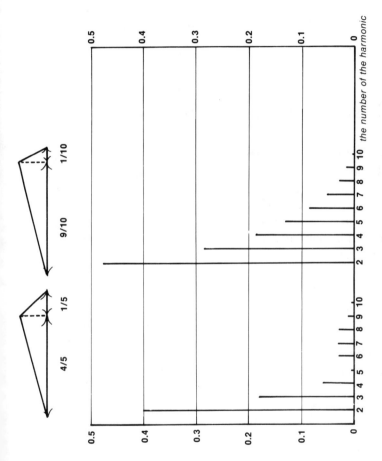

Figure 10.2B The amplitude of the Fourier components shown in Figure 10.2a for various plucking (or bowing) positions along the string. The bars at the bottom show the relative sizes of the amplitudes, with the amplitude of the lowest frequency component ("1" in Figure 10.2a) taken to be 1.0 (and not shown in the figure). The figure at the top shows in each case what the plucking position is. We see that the closer we pluck to an end of the string, the greater will be the relative importance of the higher frequency Fourier components.

void of high-frequency overtones, would be "dull." As it is, the sound, with a considerable amount of high-frequency overtones, is livelier, carries better, and attracts the listener's attention more, even though with regard to pitch it is the same tone as the one produced by bowing on the same string halfway between the ends.

The Fourier spectrum can also be influenced by the structure of the bow and how it interacts with the string, as students of string instruments know well after long hours of practicing.

The Resonant Vibrator of String Instruments

The resonant vibrator of string instruments is either a box of irregular shape or a sheet of wood, also of irregular shape. The characteristic frequencies of such irregularly shaped objects, which are two or three dimensional, form a complicated sequence of values that can be analyzed only by actually measuring them on a real vibrator or by solving complicated equations on a computer. We will do neither of these, but will simply discuss these resonant vibrators in a qualitative way.

First, it is important to understand that having a complicated resonant vibrator with many closely and irregularly placed characteristic frequencies is essential to proper functioning of string instruments. With wind instruments the musical sets of frequencies were selected by the *resonant* vibrator out of the irregular array of frequencies transmitted to it by the primary vibrator. In the case of string instruments, the situation is exactly the opposite. For them it is the *primary* vibrator that selects musical sets of frequencies, and it does so for each pitch the strings can produce. Thus the function of the resonant vibrators is more to reinforce those frequencies. As a result, for string instruments the resonant vibrator must have a great flexibility in resonating to all sorts of frequencies transmitted by the primary vibrator, that is, the strings. In fact the ideal resonant vibrator on a string instrument should be able to resonate more or less equally strongly to all the frequencies the strings transmit to it.

The irregular shape of the resonant vibrator greatly contributes to this ability. We can understand why by considering the opposite case. Let us assume that the body of a violin is a very regular rectangular box, with sides of wooden sheets of uniform thickness (see Figure 10.3). We will also assume that the primary vibrator (the strings) is attached to this box in a symmetric position, say, halfway along one of the sides. In Figure 10.3 you can see some of the characteristic vibrations of such a box, that is, of both the sides of the box and the air volume inside the box. Although these are not necessarily all the possible characteristic vibrational patterns, they represent most of them, and they can be thought of as, in each case, one dimension being active and the two others not (Figure 10.3a–c). In other words, in each picture there is variation of the amplitude along one dimension and no variation in the other two. Thus we can reduce the characteristic vibrations of such a three-dimensional box to those of a one-dimensional object, for example, of the string shown in Figure 10.1. Of course, the various vibrations in the different dimensions can also be su-

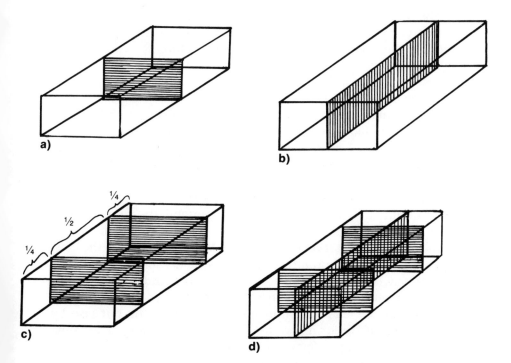

Figure 10.3 Wave patterns in a three-dimensional rectangular box. All the surfaces of the box are nodes. In addition we can have nodes along the planes shown cross-hatched.

perimposed on each other, thus forming more complicated patterns, such as that shown in Figure 10.3d. But even so, you can see that the characteristic vibrations, their patterns and frequencies, will have some regularity, and they will by no means include all the possible patterns and frequencies the primary vibrator could transmit to such a box. Thus a violin with such a regular body would have a very bad sound. It would produce some frequencies very loudly and some very softly, and hence the timbre of notes (consisting of a certain combination of such frequencies) would also be capricious. The violin depends crucially on having a body that is not regular.

In particular, the *shape* of the body of the violin is a strange one, very different from a rectangle. Its sides are made of different *materials,* of differing *thicknesses.* Finally, the components of the violin that *transmit the vibrations* from the primary vibrator to the resonant vibrator (specifically, the soundpost inside the violin) are positioned asymmetrically. Incidentally, although the external shape of the violin appears symmetric around the lengthwise axis of the instrument, the position of the soundpost and of other internal components, as well as the asymmetric thickness of the sides of the violin body, all contribute to making the violin a nonsymmetric instrument, thus adding to its ability to resonate more or less uniformly to all frequencies transmitted to it by the primary vibrator (the strings).

In the piano the resonant vibrator is the sounding board, a large wooden plank inside the piano. It is also of irregular shape, again to enhance a broad range of frequency responses.

For instruments of the violin family, it is the construction of the resonant vibrator that distinguishes the great violins from mediocre or inexpensive ones. The strings used on a violin, although their quality may differ somewhat, are roughly the same in cost, (say, within a factor of 3 or 4), and in any case constitute a very small part of the total cost of the violin. More important is the bow, which can differ in quality and price considerably and can constitute, say, one-tenth of the instrument's cost. And there are enormous differences in the quality of the body, that is, of the resonant vibrator, as one can ascertain not only by listening but also by measuring the uniformity of the frequency resonance. The range of prices one can pay for violins encompasses factors of 10,000.

Because the most crucial element in the string instrument is the resonant vibrator (since it can vary so enormously in quality), and because the scientific analysis, in detail, of such resonant vibrators (i.e., the bodies of the violin family or the sounding boards of pianos) is very complex, demanding sophistication and extensive computation facilities, until very recently such instruments were made by a trial-and-error technology and not through a scientific analysis. Now, however, our capability of understanding such string instruments scientifically has increased greatly, and it is now possible to make fairly good violins on the basis of prescriptions produced by a computer, augmented, toward the end of the manufacturing process, by a bit of trial-and-error experimentation.

The Sound Effuser

The string instruments with a three-dimensional resonant vibrator (that is, the violins, the guitar and relatives, the koto and similar instruments) produce the resonant vibrations partly in the sides of the resonating box, and partly in the air contained in the box. To get the latter out to the listener, holes are cut on one of the sides of the box. It is important to do this in such a way that (1) the hole does not spoil the ability of the box to set up air vibrations inside it, and (2) the hole emits all resonant frequencies equally well.

On guitars the hole is usually a simple circle, but on violin-type instruments it is strangely (and irregularly) shaped; it is called an f-hole because the shape is similar to that of a lowercase f. There are two such holes on these instruments.

The piano has its resonant vibrations on the sounding board and, to a lesser extent, inside the body. In any case opening the "box" by opening the top of the piano helps to get the sound out. The degree of opening can serve to regulate the loudness of the sound, for a given strength of activation of the piano keys. For example, when the piano is used to accompany another instrument that has a relatively small volume (such as a singer, or a French horn), its top is usually closed. But when it is played in a "concerto" setting, that is, the piano is the main instrument and is accompanied by an orchestra, the top is wide open to enhance the sound.

String instruments, like wind instruments, effuse sound directionally. The direction toward which the top of the piano is open receives much more sound than the opposite direction. Similarly, the direction in which the f-holes on the violin or cello point receives more sound. This is important, for example, in positioning the various players in an orchestra. Since the violins are held with the left hand and bowed with the right, the f-holes point up and somewhat to the right of the player. For this reason it is advantageous to place the violin section on the left-hand side of the stage (as viewed by the audience), facing the conductor who is in the center at the front of the stage. The celli, which effuse sound more or less straight ahead and up, can be safely placed on the right-hand side. For the same reason, in some arrangements they are placed somewhat behind the conductor (as judged by the audience) but facing the audience. The piano in the setting of a concerto is always placed so that its long dimension is perpendicular to the direction of the audience, and the pianist sits on its left-hand side, as viewed by the audience, so that the top of the piano opens toward the audience.

Summary

In string instruments the energy is supplied by arm or finger muscles, and is transmitted to the instrument by plucking with the fingers, or hitting keys with the fingers, or moving a bow that alternately catches and releases the string, thus producing vibrations.

The primary vibrator in string instruments is the string. The *frequencies* of the string vibrations are determined by *three* parameters (length, tension, and weight per unit length), even though the *wavelengths* of standing waves on such a string are determined only by the length of the string. The frequency increases with increasing tension, but decreases with increasing weight per unit length. Regulating the tension and weight per unit length can be done when building the instrument or preparing it for a performance, but not while playing. Yet the option of regulating the tension and weight lends string instruments a large range of frequencies.

The lengths of strings can be changed in two basic ways: (1) by having many strings of varying lengths, and (2) by depressing the string with a finger, thereby changing its "effective" length.

The choice of where on the string the energy is transmitted from the energy source can have a great effect on the timbre of the sound.

The resonant vibrator in string instruments is a two- or three-dimensional object of irregular shape, with many characteristic frequencies placed more or less uniformly along the frequency scale, so these resonators can readily and uniformly respond to almost any frequency the primary vibrator transmits to it.

The sound effuser in string instruments is either a hole on the resonant vibrating box or the opening of one side of this "box," such as the top of the piano. String instruments, like wind instruments, emit sound quite directionally.

Chapter Eleven
Percussion Instruments

F OR WIND INSTRUMENTS the detailed discussion of the vibrations of the *primary vibrator* is very complicated, but the vibrations of the *resonant vibrator* show a very simple pattern, in that the frequencies are determined entirely by one parameter, the length of the tube. For string instruments the detailed discussion of the vibrations of the *resonant vibrator* is complicated, and that of the vibrations of the *primary vibrator* is simpler but not too simple, in that the frequencies are determined by *three* parameters.

For the percussion instruments, nothing is simple. Both the wavelength patterns *and* the frequency patterns are complicated on both the primary and resonant vibrators. In fact the characteristic frequencies of neither of the two vibrators line up well into the regular sequence of a musical sound (that is, with frequency ratios given by simple fractions). It is for this reason that most percussion instruments have only an approximately definite pitch. Consider, for example, the cymbal, which has no definite pitch at all. Some drumlike instruments (like the timpani, which is a set of kettledrums) have a recognizable but still not very clear pitch, and the same is true for the xylophone. In spite of this apparent absence of regularity, some qualitative remarks can be made about these instruments that help to make them more understandable.

Energy Source and Energy Transfer

The energy source for percussion instruments is usually the muscles of the arm, though some drums are activated with the foot. Energy is usually transmitted by a drumstick, or some mallet or other tool used for hitting, although some drums (like those used in the music of India) are activated by hand.

The Primary Vibrator

The primary vibrator in percussion instruments is always at least two dimensional, and often three dimensional. In the case of drums, it is the two-dimensional membrane stretched over the frame of the drum. For a cymbal it is the plate of the cymbal itself, which is certainly two dimensional, and in fact deviates from a simple plane in the direction of the third dimension. The individual wooden slabs of the xylophone are definitely three dimensional, as is the tube of a chime and the metal casing of a bell. As a result the vibrational patterns are considerably more complicated than they were for a one-dimensional air column in the resonant vibrator of a wind instrument, or even than the one-dimensional primary vibrator of a string instrument, the string itself.

Furthermore, as the examples described show, the variety of percussion instruments is enormous, and hence the detailed analysis of each of them would be a horrendous task. What follows, therefore, deals mainly with two examples, the drum and the xylophone.

In the drum the primary vibrator is a circular membrane fixed at the edges along the circle. As it is two dimensional, two basic kinds of vibrations are possible (just as in the rectangular box described in Figure 10.3 there were three types of vibrations, corresponding to the three dimensions of the box). The two types are those where the lines of nodes are circles (we call these the radial modes), and those where the lines of nodes are diameters of the circle (we call these azimuthal modes from the word "azimuth" used in astronomy, which comes from the Arabic "samt," meaning "way".) We can also have superpositions of these two types. The situation is shown in Figure 11.1.

The patterns of the azimuthal modes can be described easily in a quantitative way. The lines will divide the surface into two or four or six or eight, and so on, parts, in such a way that the angle between two neighboring lines for a given mode is always the same. The radial patterns are somewhat more complicated, since the circles are not necessarily equidistant. It is, however, easy to label the radial modes by the number of circles that appear in a given mode.

When we want to relate these standing-wave patterns to characteristic frequencies associated with each mode, we encounter difficulties, since the values of the frequencies do not have the same relative relationships as the wave patterns. Figure 11.1 also indicates the values of the characteristic frequencies for each mode, relative to the simplest mode (that in the upper-left-hand corner of the array), in which the entire surface vibrates in unison, and hence there is no point (apart from the bordering circle at the edge) that has no amplitude. We can see from these numbers that they have no apparently simple relationship to each other. Furthermore, the unbracketed numbers in Figure 11.1 are for an "ideal" membrane, that is, one with zero thickness and perfect elasticity. Actual membranes behave significantly differently. Figure 11.1 also indicates (in parentheses) the actual frequencies of the modes of a real-life kettledrum head as measured in an experiment. They are significantly different from those for the "ideal" membrane.

Just as in the case of the string, we can influence what combinations of these characteristic frequencies we excite by transferring the energy to the membrane in different places, and also by using different mallets. For example, a mallet with a small, hard head transfers all the energy in a small area of the membrane (thus preferentially exciting high frequency overtones), whereas a larger, softer mallet will distribute the transferred energy over a larger area.

As already mentioned, some drums, such as the timpani, have a discernible, if not too clear, pitch. As a result these instruments can be tuned, that is, their pitch can be changed by changing the tension of the membrane (just as changing the tension on the violin string changes its pitch). On some timpani this can be done relatively easily with a foot pedal, and so a given musical note can be made to "slide" in pitch while it resounds. This is used very effectively, for example, in the unique composition by the 20th cen-

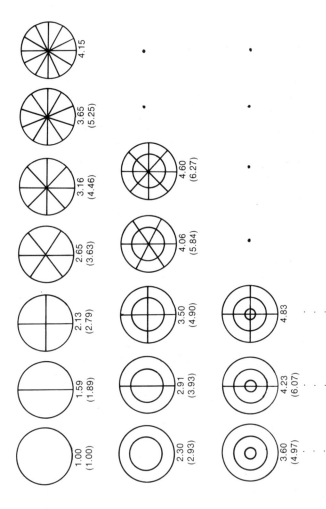

Figure 11.1   Vibrational modes for a circular drumhead. The rim of the circle is always a node, since the drumhead is clamped down along its circumference. In addition we can have nodal lines on the surface of the drumhead. Two areas separated by such a nodal line vibrate 180 degrees out of phase with each other. There are radial nodal lines (circles) and azimuthal nodal lines (diameters of the circle). The numbers below the figures indicate the ratio of the frequency of that mode with respect to the frequency of the simplest vibration (shown in the upper left corner). The unbracketed numbers are calculated for an idealized drumhead, and the bracketed numbers are for an actual drumhead as measured.

tury composer Béla Bartók, "Sonata for Two Pianos and Percussion."

Let us now turn to our second example, the xylophone. That instrument consists of a row of wooden slabs of varying lengths, each of which is attached to an overall frame at two places, but *not* at the ends. The instrument is played with two mallets, which hit the wooden slabs.

These slabs are the primary vibrators. Compared with, say, a violin string, they are more complicated to analyze for at least two reasons: (1) They cannot be considered one dimensional even as an approximation because their length is only three to four times their width or thickness. (2) They are not fixed at the ends but at some intermediate point (as in the xylophone) or hang freely (as in the glockenspiel). For these reasons their vibrational patterns are more complex; see Figure 11.2. Furthermore, just as was the case for overly thick strings, the ratios of the frequencies will strongly deviate from what we have for the (idealized) thin string of the violin. The frequencies are also indicated in Figure 11.2, relative to the frequency of the simplest mode. Again we see that the ratios are not very close to the simple fractions that make the sound musical.

The Resonant Vibrator

In the case of the drum, the resonant vibrator is the air volume inside the drum, that is, between the top and bottom membranes of the drum, or between the top membrane and the bottom metal structure in the case of the timpani. Although the shape of that air volume is somewhat regular (for example, the shape of a cylinder for a snare drum, or the shape of a wine glass for the timpani), and hence the patterns are not quite as varied as in the body of the very irregularly and asymmetrically shaped violin, the quantitative analysis of the vibration patterns still requires sophisticated mathematics using computers.

In some percussion instruments, the situation in simpler. Consider, for example, the chime, which consists of a series of metal pipes of differing lengths. We hit these pipes from the outside with a hammer, and the pipes themselves form the primary vibrator.

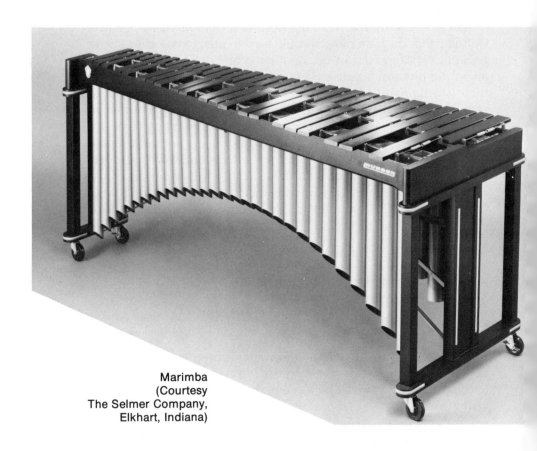

Marimba
(Courtesy
The Selmer Company,
Elkhart, Indiana)

Timpani
(Courtesy
The Selmer Company,
Elkhart, Indiana)

The author striking a gong in a monastery in Southeastern Thailand. The gong, made of metal, is some 10 feet in diameter and has a thickness of half an inch or more, so that its fundamental frequency is so low that a hand placed on the gong can feel the individual vibrations. Such an emphasis on low frequencies enables the gong to be heard at quite large distances.
(Photo courtesy of Dr. Kopr Kritayakirana)

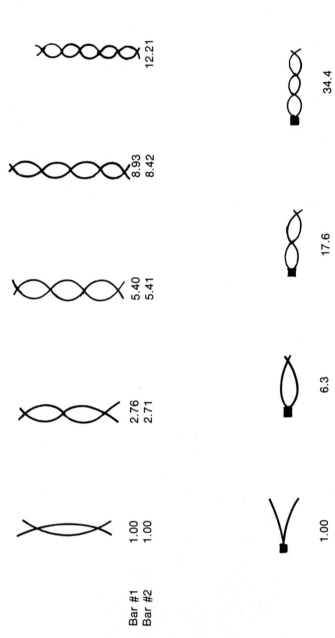

Figure 11.2 Vibrational modes and frequency ratios of two different steel bars freely hanging, and one steel bar clamped at one end.

The resonant vibrator is the air column inside the pipe, which then behaves almost like the resonant vibrator of wind instruments, which, as we recall, was very simple. To be sure, the fact that the envelope of this tube also vibrates and thus changes the shape of the tube slightly also affects the type of vibrations that are characteristic of the air column inside the pipe, but this is not a large effect. Thus chimes do have recognizable pitches, not because the vibration patterns of the primary vibrator (the pipe itself) are simple and follow the musical recipe (since that is not the case), but because the resonant vibrator is simple and conforms to the frequency pattern of musical sound.

A similar situation exists with the marimba, in which special resonating pipes are affixed to the instrument to serve as resonant vibrators.

A peculiar percussion instrument is the triangle, which is a metal rod shaped into an equilateral triangle, and held suspended on a small chain. The instrument is then hit with another metal rod. Here the triangular rod is the primary vibrator, which directly transmits the vibrations to the air around it, without the services of a resonant vibrator. As a result the total loudness of the sound emitted by the triangle is very small. Yet the triangle plays an important part in the overall orchestral sound, since its sound has relatively strong Fourier components at extremely high frequencies, and hence it stands out among all the other sounds, which are at lower frequencies. A given triangle can produce only one type of sound (its "pitch" is hardly, if at all, recognizable), and so it cannot be tuned or its frequency changed in any way.

We might ask what function some of these percussion instruments have when they cannot even produce a recognizable pitch. The answer is that it is exactly because they do not emphasize pitch that they can be dominant in defining rhythmic patterns. Indeed, ethnic music that utilizes predominantly percussion instruments will also be strong on rhythm, while other cultures in which wind or string instruments play the most important role will highlight melodic patterns.

On rare occasions an instrument of a given type can be used as another type. For example, in some modern compositions the violin is briefly used as a percussion instrument as the player gently taps on the violins body instead of playing on the strings. Such a strange use of instruments is, however, only employed for temporary special effects.

Summary

In percussion instruments both the primary vibrator and the resonant vibrator are complicated to analyze since they are two or three dimensional. One can qualitatively indicate the standing-wave patterns on the simpler of these instruments, but the corresponding frequencies form a more irregular array that does not conform to the frequency ratios that make a musical sound. As a result most percussion instruments have either no recognizable pitch, or only a somewhat uncertain one. Thus percussion instruments are generally not so helpful in creating the melodic structure of music, but are extremely well suited to emphasizing rhythmic patterns.

Chapter Twelve
The Human Voice

I‌T IS OFTEN said that the human voice is the most complex and sophisticated instrument. This is true, but not for the reasons some people might like to imply. These people would suggest that the human voice is the most "natural," most "direct" instrument; that it is the only "human" instrument; that it is most completely and directly under our control; and hence that it must be the greatest. In fact, however, all instruments are humanly activated, and as we shall see, most other instruments are much more under the conscious control of the player than is the human voice.

Two Special Features

In many ways it is true that the human voice is the most versatile and sophisticated instrument. From our scientific point of view, this is so because the voice has two features almost completely absent in other instruments, and in particular in the ones we have discussed so far. These features are:

1. Both the primary vibrator and the resonant vibrator can change their sizes and shapes while the music is being performed.

197

2. It is possible to vary the effuser drastically by placing obstacles in the path of the air streaming out.

These two additional features create an enormously expanded range of possibilities for the human voice. Largely because of the first of these, there is an infinite variety in the character and timbre of human voices. Indeed, probably no two human beings out of the many billions born throughout history have ever had identical voices—a fact that is utilized in the method of identifying people by their "voiceprints," that is, by a scientific analysis of their voices along the parameters we have been discussing.

The same feature also allows us to recognize singers from records even without the use of scientific measuring instruments.

The second of the two features has an even greater impact: It allows us to use our voice for speech, in addition to using it for making musical sounds. In fact we can even combine the two functions and make musical sound while speaking, as we do when we sing the lyrics of a song.

Having acknowledged how special an instrument the human voice is, we recognize that basically it is a wind instrument, and so we will not need to learn about special new structures but simply refer to the general characteristics of wind instruments.

Energy Input and the Primary Vibrator

The energy for the human voice comes from the abdominal and chest muscles that extend and contract our lungs. We thus inhale air into the lungs and then exhale it to create an airflow—just as with any wind instrument. However, there is one difference: unlike all other instruments, in the case of the human voice *all parts of the instrument are parts of one's own body,* and, in fact, mostly parts *inside* the body.

In particular, the primary vibrator is the vocal cord, which is a fleshy and muscular layer of tissue attached at the edges to the larynx; it hangs freely into the larynx, and has a slit through which air can pass. Because this slit separates the two parts, we sometimes speak of vocal cords—which should remind you of a double-reed instrument, such as the oboe. Indeed it functions very much as does the reed in an oboe, producing vibrations because of the Bernouilli effect.

The big difference is, however, that while the reed must be manufactured in its final form before the performance starts, the vocal cord can be altered considerably by the action of one's muscles—both by flexing such muscles and by relaxing them. Thus the characteristic frequencies of the vibrations of the vocal cords can be regulated, at least within certain limits. For example, the pitch of the sound we make can be varied to a considerable extent, which is easily two octaves and can even be more, without either opening or closing side holes on our necks or noses, or elongating and shortening our necks or heads, or opening and closing secret bypass passages inside our heads—as would be true of traditional wind instruments. Instead pitch is regulated by muscular control of the vocal cords and resonant cavities. This way of creating a range in pitch is somewhat limited compared with those used in traditional instruments, and no human can compete with the seven or more octaves of the piano. But compared with wind instruments, the human voice does quite well.

This variability of the geometrical parameters and of the stiffness of the primary vibrator carries over also to the resonant vibrator, which in the human voice consists of the rest of the larynx and the oral and nasal cavities in the head. The size and shape of these cavities and the passages among them can be regulated, both consciously and automatically. As we remember from the discussion of wind instruments, the characteristic frequencies of the resonant vibrator play a dominant role in determining the Fourier spectrum of the sound produced by the instrument. This is also true for the human voice.

You can easily experiment with this, even if you are not a trained singer. Sing a given note (pitch), and then try to change its "vocal color": Sing it serenely and warmly, then try the same note using a nasty, sarcastic tone. The latter will have many more high-frequency overtones mixed in with it. It will be a "whiter," a more penetrating sound. These expressions are, of course, symbolic, as is most of the vocabulary used by music critics to describe performances.

Voice—An "Internal" Instrument

This brings us to a peculiar and very interesting aspect of the human voice as a musical instrument, briefly referred to earlier.

For all other instruments, sound is produced through an interaction of some external parts of the human body (lips, fingers, arm, feet,) with a mechanical device. As this interaction can be observed by both the player of the instrument and the musician's teacher, the instruction can be explicit in terms of advice on what to do with the relevant parts of the body. The student, depending on the instrument involved, is told what to do with the fingers, how to hold both hands and arms, how to fit his or her lips to the mouthpiece, and so on. At least the purely technical (as distinct from musical) aspects of instrument playing therefore can be handled in a rational way.

The situation is quite different when one studies voice. Many volumes have been written on singing techniques, and thousands of voice teachers have lectured their pupils on how to sing, but all of these have had to use, by necessity, figurative, symbolic language. To be sure, some general advice about relaxation, about using the diaphragm muscles for breathing, about trying to keep the tongue out of the way, has objective meanings also. But exhortations such as "open your passages as if you were biting into an apple," "open up as if you were about to vomit," "sing into your forehead," or "smile inside your head" represent poor substitutes in an effort to convey a certain overall state of the inner muscles of the body over which we have at best partial control.

How You Hear Yourself

Adding to this frustrating situation is another feature unique to singing as compared with other instruments. As a singer one hears oneself in a way that is fundamentally different from the way the audience does. The audience receives the sound signal entirely through the vibrations generated in the air, whereas in the singer some of the auditory stimulus is conducted to the ear through the singer's own bones. Since these two ways of transferring sound have quite different relative efficiencies at various frequencies, the overall timbre of the sound will be quite different. You have probably experienced this when you have listened to your own voice, as on tape or through a public address system. It is easy to blame the "sound of a stranger" on "poor electronics," but this is only partly justified. The major effect comes from the fact that you hear yourself differently from the way others hear you. This is one of the main reasons why even the most accomplished singers have to defer to the opinion of coaches and voice teachers as to "how they sound," whereas no concert violinist would have to do such a thing. To the violinist playing sounds almost exactly the same as it does to someone else standing nearby.

We have not yet mentioned the effuser in the instrument of the human voice, namely, the mouth and nose. This is of great importance in singing, since here, more than anywhere else, the sound must get out of the body of the performer if it is to be heard. Singing teachers have similes for this also: "Sing as if you were smelling a flower" (i.e., open your passages so the sound can come out). "Make sure the shape of your mouth is a square." And so on. In any case the efficiency with which the sound is emitted can make all the difference to the singer who must sing hour after hour in an enormous hall and be heard over the sound of other singers, the chorus, and a large orchestras.

Singing "Over" the Orchestra

Another interesting observation is in order. One might wonder how, say, one tenor can be heard over all the other musicians. Does the tenor actually sing "louder"? The answer lies in the ambiguity of the term "louder," which we can now understand on the basis of the physics we have acquired. If the question is whether the tenor produces more total sound energy than all the other musicians together, the answer is clearly "no." Total sound energy means the squares of the amplitudes of sound vibrations summed over all the frequencies at which sound is emitted. The tenor, however, does not need to overpower in energy the sum of all other musicians *at all frequencies*. It is sufficient to be dominant in a certain frequency range. For example, an operatic tenor such as Luciano Pavarotti (whose voice has an unusually large admixture of high-frequency overtones, even for a tenor, which causes the extrordinary "brilliance" of his voice) would outdo the combined forces of other musicians in the frequency range around 2000–3000 vibrations per second (see Figure 12.1). The ear of the listener, which, as we know, can analyze each frequency region separately and then synthesize the information in the brain, will then easily

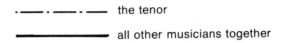

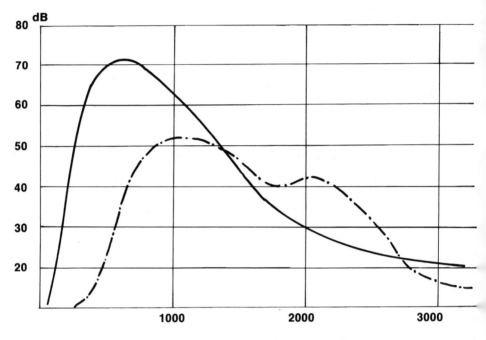

frequency (vibrations per second)

Figure 12.1 Loudness distribution for various frequencies for an orchestra and chorus (solid line) and a tenor soloist singing with the orchestra and chorus. The tenor emits much less sound energy than the rest of the musicians, but is louder in a limited frequency range around 2000 per second. This is sufficient for the audience to hear the tenor singing "above" the orchestra and chorus.

recognize Pavarotti's voice among the myriad of sound impressions even though Pavarotti may put out only 1 percent of the total sound energy heard. It is, of course, easier for the higher voices (soprano or tenor) than the lower voices (alto or bass) to find such a "frequency window" where they can excel.

The same principle is also utilized by the manufacturers of some contemporary telephones and also of quartz wristwatches, which (often quite annoyingly in public places) squeak every hour on the hour. They use very-high-frequency sound, which is not likely to occur in the natural sounds around us, and therefore a relatively very low-intensity sound of that unusual frequency will catch our attention. The "chirping" phones, for example, use much less power to ring than did the old, conventional phones, which employ a bell that rings at moderate frequencies.

These high-frequency admixtures in a voice come primarily from the appropriate use of resonant vibrators and not so much from the primary vibrator, that is, from the vocal cords. To be sure the vocal cords themselves will also vary in size and shape from person to person. Our general rule holds there also—namely, larger people with larger vocal cords will *tend* to produce lower frequencies. It is no accident that, generally speaking, bass singers are larger than tenors. Small children have relatively high-pitched voices. The sound a tiny cicada is at much higher frequencies than the sound of a huge elephant. The difference in pitch between male and female singers, however, does not arise from the effect of size. There the different structure of the larynx (which in males develops during puberty) plays a role, but the extra admixtures at high frequencies come from the use of head cavities.

The Helium-Filled Singer

The importance of the resonant vibrator (the head cavities, etc.) in determining the pitch of spoken or sung sound can be demonstrated in an amusing way by asking a singer to inhale helium gas and then to speak or sing. Helium gas is harmless (although, of course, existence in *pure* helium gas for an extended period results in suffocation due to a lack of oxygen). In any case, the speed of sound in helium is about three times greater than in air, which is important for our experiment.

The singer who has inhaled helium gas will be able to produce only very-high-frequency squeaks—"Donald Duck speech." One's vocal cords in the presence of helium will work just as they always do, since the vibrational frequencies of the cords depend only on their own geometrical parameters and on their stiffness. Thus the primary vibrator will transmit the same mixture of frequencies as usual. The resonant vibrator, however, is a cavity, and its size determines the *wavelengths* of the standing waves of the gas that fills the cavity. Therefore, with helium inside the singer, the frequencies produced by the primary vibrator will be as before, and so will the *wavelengths* of the characteristic standing waves in the resonant vibrator. The *frequencies* of the resonant vibrators, however, will be given by the speed of sound in the gas divided by the characteristic wavelengths of the resonant vibrators, and since the speed of sound is now three times larger than it was in air, the frequencies will also be three times larger.

If the pitch and character of the human voice were primarily determined by the frequency of the vocal cords, the voice would not change whether air or helium are in the singer. If the main determinant is, however, the resonant vibrator, the frequency should rise by a factor of three or by an octave and a fifth. When we carry out our experiment, we find the latter to be the case, thus demonstrating the crucial role played by the cavities in shaping the pitch and timbre of human voice.

Vibrato

At least in Western music, the trained singing voice has an important characteristic contributing to its perceived attractiveness. It is called the "vibrato," and it refers to a periodic change in intensity, three to eight times a second. Singing that does not feature this is sometimes called a "straight" sound, and it is characteristic of children's singing voices (as well as those of untrained adults). In some Western music, such as that of the Renaissance, performance practices dictate the use of such straight voices. At the other extreme, some popular singers cultivate such vibrato since it has connotations of emotionalism, warmth, sexiness, and so on. Our taste for such vibrato demands a definite frequency range for the intensity variations: less than two a second will be judged unpleasant, and more than eight or so a second will have little effect. Within even this rather narrow range, the effects can vary. Some varitones, like Emilio deGogorza, have a relatively fast vibrato (perhaps six per second), which contributes to the individualistic timbre of their voices. Another example of a voice with a fast vibrato is Joan Baez.

Vibrato, which is (primarily) a variation in intensity, is different from "tremolo" or "wobble," which is a variation in *pitch* in the singing voice. This is a tolerable, and perhaps even normal, part of the singing process if it is kept to a small range of changes in the frequencies, say, 1–3%. If, however, the range of the change in frequency is much larger, it is perceived as definitely ugly. Low voices (altos, baritones, and basses) are particularly afflicted with this defect, which is often developed later in the singer's career. The frequency of the frequency change also matters in this respect: large frequency variations that occur with relatively low frequency (one to two per second) are considered particularly objectionable.

Vowels

We now turn to the other mode of vocal communication, that is, to speech. We produce two kinds of sounds when speaking. One, much closer to musical sound, is called vowel. When we speak a vowel, the process is, in most respects, like singing, except that we regulate our resonant vibrators not to produce the musical overtones needed in the Fourier spectrum of a musical sound, but in a less special way that produces a less regular set of overtones, and so the resulting effect is closer to noise. The effusion of the sound, however, is the same as in singing; namely, we place no obstacles in the path of the sound.

You can easily experiment with vowel production, either by speaking or by singing vowels such as the "a" in "law" and that in "case"; the "e" in "met"; the "i" in "lit"; the "o" in "hold" and that in "hot"; the "u" in "moot," and that in "hut"; the "u" in the French "dur," and that in the German "Brüder"; "ö" in the French "fleur," and in the German "Möwe." You will see that you can sound any of these vowels continuously, that they leave your mouth without obstruction. You will also find that the differences between these vowels is in the way your lips and mouth form them—or at least these are the elements of which you become consciously aware.

You also soon find, when experimenting with vowel production, that there is really a continuous range of vowels. Start with an "i" as in "meat," and alter your mouth and lips slowly to change over to an "e" as in "care," then on to an "a" as in "tar," followed by an "a" as in "mall," and then to "o" as in "hold," and finally ending up with "u" as in "lute." This chain of transitions can be made gradually, and these "road signs" are simply six stages in a slow evolution that takes you through scores or hundreds of different vowels.

Incidentally, Isaac Newton, probably the most original scientist to date, when still a child, experimented with sound and remarked that when a bottle is gradually filled with water, the sound it makes goes through a continuum of vowels from "u" to "i." Try it yourself.

Indeed, different languages use different vowels. Each language chooses a discrete set out of this continuously infinite set of vowel possibilities, and postulate those to be used in that language. Some languages, including English, go beyond this and specify "diphthong" vowels, that is, vowellike sounds that start out on one of the "pure" vowels and slide over into another. For example, "howl" can be described as going through three of the "pure" vowels of, say, the Italian language: It starts on something like the "a" in "caro," continues to something like the "o" in "Roma," and ends up on an "u" that is something like in "uscita."

No wonder that people who have grown up speaking one language have difficulty in acquiring another language with its completely different sets of vowels. People's various "foreign" accents are largely determined by the different types of vowels in different languages.

In some languages it is not only the type of vowel but also its length that plays an important role. In English the short "i" (as in "ship") and the longer "i" (as in "sheep") not only are different in length but also somewhat in timbre: The "i" in "ship" is less open and has an element of the German "ü" (e.g., "Brüder") mixed in it. In contrast the two vowels in Hungarian, one denoted by "i" and the other by "í," are different only in length.

To make things even more complicated, in some languages not only the type and length of the vowel are relevant, but also the intonation, that is, the way the voice rises and falls in pitch as it proceeds through a word. Some Asian languages are famous for this, such as Chinese and Thai.

As mentioned, vowels can be either spoken or sung. The same word, however, when spoken, may contain certain vowels that are slightly changed when sung. This is so because the singer also must obey the demands of the musical expression, and needs to pour out musical sound in the appropriate amounts and quality.

As an example, take a bass singer who has to sing the German word "Mut" on an F-sharp just below the 440 A. That note is a very high one for a bass, close to the end of his range, and singing such a "low" vowel (like that "u") on such a high note can be very difficult. He will therefore change the vowel a bit, not so much that the word becomes unrecognizable, but enough to ease the technical difficulties. In particular, some "higher" vowels, such as "ö" or "ü," will be admixed. The same problem arises when one sings the nasal vowels of the French language.

Consonants

The other type of sound we make when speaking is called a consonant. In making it some kind of obstruction is used in the effuser of the voice instrument, that is, the mouth and the lips. The obstacle can be formed by the lips, the teeth, and the tongue and palate in combination. There is also the "noisy" but unimpeded exhaling of air that results in what we call the consonant "h," in which the vocal cords play no role. The nature of the obstacle can also be of several types: we might close the mouth altogether and let the air exit through the nose (for example "m" or "n"). Alternatively, we might close the exit and then let the air "break through it" suddenly (for example, "t" or "k"). Or we might just constrict (but not completely close) the airflow, causing friction and turbulence (for example, "f" or "z"). In each case the consonant might be more silent or louder ("voiced"). For example, the difference between "p" and "b" can be said to be that the first is "unvoiced" and the second "voiced."

If you have tried these examples and remain less than totally convinced, you have company. I also consider these classifications a valiant but not altogether successful effort by scholars to convert a continuous range of possibilities, for both vowels and consonants, into some discrete classification, since this is easier for people to understand and use. The international phonetic alphabet is the result of such efforts—a list of characters that presumably allows one at least to approximate each sound used in each language. The task is an impossible one to carry out exactly, especially for vowels, but such efforts help linguists, speech therapists, singing teachers, manufacturers of telephones and of talking robots, music critics, and a miscellany of other professionals.

Summary

The human voice as an instrument has two special features: both the primary and resonant vibrators can change their sizes and shapes; and the instruments can be altered by placing obstacles in the path of the air in the effuser, thus allowing both singing and speech. The energy is supplied by the abdominal and chest muscles, and the energy is transferred as a moving airstream. The primary vibrator is the vocal cords, and the resonant vibrators consist of the various cavities in the throat and the head. All parts of the sound production take place inside the human body, and hence teaching somebody to sing is a complicated process, the more so because singers hear themselves differently from the way others hear them. A singer can "come through" over massive orchestral sound by being able to outpower the competitors in a particular high-frequency range. This high-frequency "edge" of the singer originates mainly in the resonant vibrators, as an experiment with a helium-filled singer can easily demonstrate. Vibrato is primarily a periodic intensity variation in the singer's voice. A pleasant effect is created if the frequency of the variation is between three and eight per second. The tremolo, on the other hand, is pitch variation and generally is considered in a negative light. Speech consists of two kinds of sounds: vowels and consonants. Vowels pass through the system without obstruction, except in the vocal cords. There is one big continuum of vowels, out of which each language selects a discrete set for its own use. The length of vowels and intonation also count in speech. Consonants are formed through various types of obstructions in the path of the emerging airstream. The various classifications used there are also an attempt to classify a continuum.

Part V
Consuming Music

Chapter Thirteen
The Environment for Listening

What Is the Problem?

Music, FROM THE listener's point of view, has many functions. In some of these music is one of several elements and is used to heighten an overall experience or to serve as a background to it. The objective of church music is to enhance a spiritual involvement, military marches are intended to augment the martial spirit, soft background music in supermarkets is supposed to soothe the harried shopper, movie music is designed to underscore the story, one's stereo or radio playing softly while one is doing something else is meant only as a background sound for company, music for ballet serves to mark time and to add to the visual impression, and so forth. In other situations, however, music becomes the primary, or even sole, preoccupation of the listener, although the mind may wander—recalling, associating, and forming personal memories, images, and thoughts stimulated by the music. In all these functions, and particularly in the last one, the listener's environment is of great importance.

Some of this environment relates to the psychological atmosphere. The solemn confines of a church, the informal setting of a jazz festival, the collective and relaxed atmosphere of a concert hall, or the comfortable privacy of one's living room contribute to the listener's psychological well being. This essential aspect of listening to music is outside the confines of our subject matter in this book. Instead we are concerned with that part of the environment that tries to ensure that the sound reaching the listener's ear is ideal for maximal enjoyment.

This objective is important not only for the listener. We might say that three major elements contribute to the enjoyment of music. First, is the composer who created the sound patterns to be performed. Second, is the artist or artists with their instruments, who interpret the composer's sound patterns and recreate them. Finally, there is the listener who perceives, comprehends, and absorbs the music. At each stage of transition in this chain, shortcomings and frustrations of the transfer process can plague the participants. In particular, an artist who renders a composition in a way the artist considers superb may be thwarted by the acoustic shortcomings of the room or hall where the performance is held. Also, to judge the performance, the artist must rely to some extent on his or her own hearing of the music, and this may also be distorted by the environment. It is common, for instance, in high school auditoriums (which are designed to be used for various purposes and hence are by no means ideal for musical performances) to have a high, roofless stage with velvet curtains at the back, an arrangement that results in dulling, absorbing, and channeling away the sound to unwanted locales, thus confusing and annoying both the performers and the audience.

There is a long list of requirements from the point of view of the listener for an optimal listening experience. The music must not be interfered with by extraneous noises from outside or inside the performing area. The sound must be of a sufficiently high volume for everybody in the audience. The sound should have clarity, that is, the pitches and rhythmic patterns should be well distinguishable and "lively." The level and quality of the sound need to be uniform throughout the listening area and throughout the ranges of frequency and volume of the music. The listener wants to have the feeling of being "enveloped" in sound as if in the midst of the performers. At the same time, the listener wants to have some spatial perception and differentiation if faced with a large group of performers. In most situations the listener does not want to hear an echo, that is, a second arrival of the sound separated in time from the first one. On the other hand, the listener is sensitive to the reverberation patterns of the sound, that is, the way the sound "dies out" after the initial arrival, and does not like a total absence of reverberation. This list of requirements is a very severe one to

fulfill. Although scientific considerations such as those we will discuss (and more sophisticated ones) help considerably in creating a
great listening environment, efforts to do so remain a combination
of systematic science and trial-and-error technology, and as a
result conspicuous failures by noted practitioners can be found
along with their successes.

The difficulty of the task is compounded by the large variety of
music performed. A guitarist giving a solo recital, a popular singer,
a rock band, a string quartet, a 100-piece orchestra with a chorus
of 300, a jazz band, an operatic performance, a musical, and a dis-
play of electronic music all have very different circumstances for
which the ideal environment varies equally widely.

Reflections

Virtually all listening to music occurs in an environment that is
spatially limited by walls (or other objects) and so the sound
reaches the listener not only directly, but also after reflection from
the walls. The following is a discussion of the various elements
that affect the reflected sound in such listening spaces.

We already know many of the important aspects of this problem. You will recall that we learned that waves, when impinging on
a flat surface, reflect back at the same angle (but on the other side
of the perpendicular to the surface) as the one at which they im-
pinged. We also learned that when waves impinge on such a surface, only some of the incident wave is reflected, while the rest
passes through the boundary represented by the surface, and
penetrates into the medium that makes up the surface. This refracted part of the wave continues at an angle that is different from
the angle of incidence. Finally, we also learned that when a wave
travels through a medium, some of it is absorbed in that medium
(and its energy turns into heat), so that there is a damping of the
wave. These simple elements of the behavior of waves are almost
enough for the discussions in this chapter.

When considering the reflecting qualities of a certain wall or
other object, it is useful to characterize the material of the surface
by a number indicating the percentage of wave intensity (i.e., energy) that is absorbed (called the absorption coefficient) and an

associated number that indicates the percentage of wave intensity (i.e., energy) that is reflected (reflection coefficient). Since energy must be conserved, and can go only into reflection or absorption, the two percentages must add up to 100 percent. The actual values of these coefficients denote fractions rather than percentages. For example, a coefficient of 0.63 corresponds to 63%. Note that in this context, "absorption" means any wave that is not reflected; that is, it means the wave that penetrates the surface and refracts. Whether that refracted wave goes through the wall, or literally is absorbed and its energy turned into heat, is, in most cases involving acoustics, of no interest, and hence we call any wave not reflected an absorbed wave, even though this terminology is a bit sloppy.

The absorption (and reflection) coeficients of various materials range widely, from almost 0 to almost 1.00 (corresponding to 100 percent). Furthermore, these coefficients vary, for the same material, with the frequency of the wave. Figure 13.1 gives the absorption coefficients for some materials relevant to listening environments for several frequencies of the waves. You can easily calculate the reflection coefficients from the illustration by subtracting each number from 1.00.

Note that the graph in Figure 13.1 is only approximate: the values of the absorption coefficients were taken from a table at six points, and then those points were connected by straight lines, thus creating apparent kinks at the six points that do not exist. Thus the figure should be used only for a rough orientation concerning the size and frequency variation of the coefficients for various materials.

With this knowledge we can easily describe what happens to a particular sound wave in a hall (see Figure 13.2). It is generated by some source, say, on the stage. From the source sound waves will travel in all directions, but for the moment let us follow only that part of this wave that travels in one particular direction from the source. We will also assume that the wall of the hall is made of a material with an absorption coefficient of 0.5. This is not unreasonable; it corresponds to fairly heavy drapery. In Figure 13.2 the path of the sound wave is traced for the first 10 reflections, and next to each path segment the sound intensity is indicated, assuming that the original wave emanating from the source in this direction had an intensity of 1.00.

We see from the figure that after the 10th reflection, the original intensity has dropped to about 1/1000 of its original value, or by 30 dB.

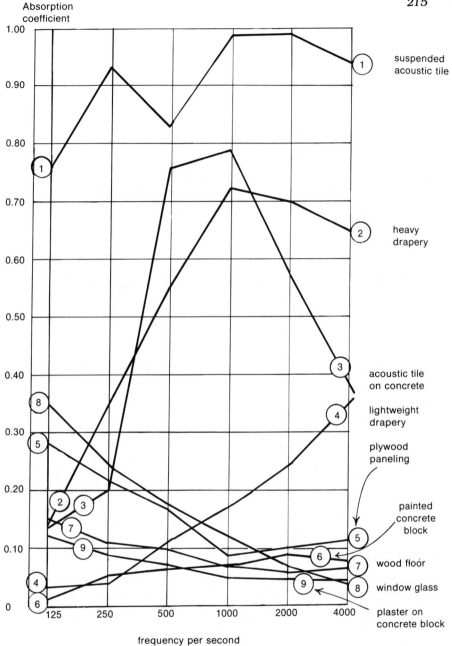

Figure 13.1 The absorption coefficient for various materials used in acoustics, at various frequencies. Data were taken at a few frequencies and indicated by points, and then the points were joined by straight lines. Hence the values at frequencies between the measured points are only approximate. One sees that, for most materials, the absorption coefficient is quite small (below 0.30). For such situations the approximation for the reverberation time explained in the text is justified.

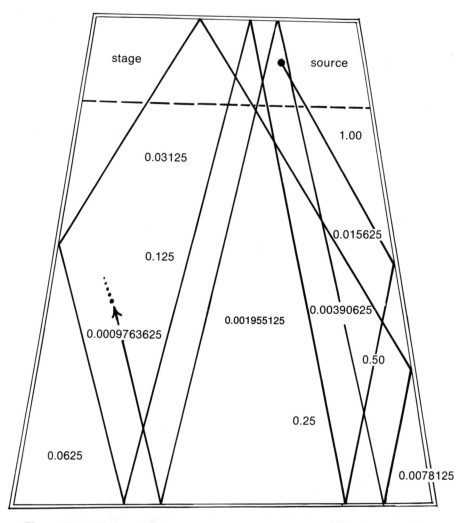

Figure 13.2   The intensity of reflected sound after repeated reflections from a wall with an absorption coefficient of 0.5. The numbers next to the path segments indicate the sound energy as compared with the original sound wave starting from the source, the energy of which is taken to be 1.00.

How long did all this take? Let us take a reasonable size for the hall, say, 200 feet long and 100 feet wide. As a rough estimate, we see that the sound we are tracing crossed the hall lengthwise seven times, and at the same time also crossed it crosswise about three times, so the total path length is roughly (7 × 200) plus (3 × 100), or 1700 feet. Since sound travels about 1000 ft/s, the whole event up to the 10th reflection takes about 1.7 seconds. You can now easily convince yourself that if we were to let the sound bounce around 10 more times, its intensity would drop by another factor of 1000 (or by another 30 dB), and this additional travel would take another 1.7 seconds. So the drop from the original intensity to 60 dB below that would take 3.4 seconds.

This length of time, of course, depends on the initial direction from the source in which the sound traveled. If you look at Figure 13.3, you will see another sound wave after 10 reflections. In this case the total path length is about 7½ times a crosswise traversal plus 2½ times a lengthwise traversal, or a total of 2½ times 200 feet plus 7½ times 100 feet, or only 1250 feet. Thus with this pattern of path the sound would drop 30 dB in 1.25 seconds and not 1.7 seconds, as was the case for the other path length shown in Figure 13.1.

Reverberation Time

In reality sounds emanate from the source in all directions, and so there will be a huge array of various reflection patterns. Some will attenuate faster and others more slowly, since some will have more reflections in a given path length, and others fewer. Thus the original sound coming from the source, as heard from a particular spot in the hall, will also attentuate, as a result of the various attenuating sound waves reaching that spot at various times. The combined effect can be characterized by a number that gives the number of seconds in which the sound perceived at a given spot in the auditorium decreases its intensity by 60 dB. The 60 here is merely a convention, to give this number (called the reverberation time) a unique definition.

From the foregoing it is clear that the reverberation time will depend on the dimensions of the hall (which will make the path

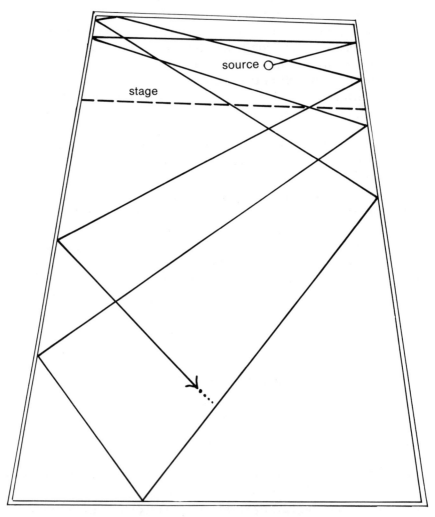

Figure 13.3 A different arrangement of path segments starting from the same source as in Figure 13.2. We see that in this arrangement the average path length between reflections is shorter than it was in Figure 13.2.

lengths between reflections larger or smaller) and on the reflection (or absorption) coefficient of the wall, which will take a smaller or larger bite of the incoming sound wave energy during one reflection. These two dependences will now be discussed separately.

Dependence on Size

Let us start with the dependence on the dimensions of the hall. That there is such a dependence can be easily understood by considering our example in Figure 13.2. Let us change the dimensions of the hall so that both length and width are increased by a factor of three. That would mean that the sound would take three times as long to cover each path segment, and so the 10th reflection would be reached not after 1.7 seconds, but after 5.1 seconds. In other words, we see from the figure that the reverberation time is proportional to a one-dimensional quantity (the path segment length), and hence will scale just like the one-dimensional size (length or width) of the hall. The powerful technique of scaling comes to our aid again and yields a simple result.

Note that in Figure 13.2 we drew only two dimensions of the three-dimensional hall, and also drew the path only in two dimensions. Putting in the third dimension, however, does not change the result. You will see this easily, if you visualize the sound wave bouncing back and forth in a three-dimensional volume and then observe that scaling up that volume by a given factor in each of the three dimensions will change the length of each path segment only by that factor itself.

Scaling, however, does not help us if we want to compare the reverberation times of two halls of different sizes *and shapes.* We can easily see that the shape will make a big difference. Consider, for example, an absurd but conceivable hall 200 feet long but only 20 feet wide and 20 feet high, just like a tunnel. It is obvious that in this low, narrow, but long tunnel sound waves will reflect much more often than in a hall of more regular dimensions. Look at Figure 13.4, which compares (in two dimensions) two halls. Let us assume that both have the same height, but one is 100 by 100 feet (Figure 13.4a) and the other is 400 by 25 feet (Figure 13.4b). Thus the two halls have the same floor area and also the same volume. Both of these figures show the path of a sound wave starting from the same (arbitrary) point and going initially in the same (arbitrary) direction. In both figures only the first 1000 feet of path are indicated. We see, however, that in Figure 13.4a the 1000 feet of path length involve only 13 reflections whereas in Figure 13.4b, 36 reflections are involved. Thus the reverberation time clearly will be much shorter in Figure 13.4b than in Figure 13.4a. We saw earlier that twice as many reflections in a given time period mean half as long a reverberation time, and so we conclude that if we increase the area of the wall and thus get twice as many reflections in a given time period, the reverberation time will be halved.

These observations allow us to draw a final conclusion as to how the reverberation time depends on the dimensions of the hall. We saw that the reverberation time is proportional to one dimension of length and that it is inversely proportional to the wall area. To satisfy these two conditions, the reverberation time must be proportional to the cube of length and inversely proportional to the wall area (which is the square of a length), because the overall dependence is 3 − 2 = 1 dimension of length. The only quantity we have that describes the hall and which is cubic in length is its volume, so that we can finally say that the reverberation time will be proportional to the volume of the hall divided by the wall area.

Dependence on Absorption Coefficient

We now turn to the other quantity on which the reveration time can depend, namely, the absorption (or reflection) coefficient of the wall. We can see from the examples we discussed that the larger the absorption coefficient of the wall, the smaller is the reverberation time. For example, a direct calculation will show that if the absorption coefficient of the wall in Figure 13.1 is not 0.5 but 0.4, then the intensity after the 10th reflection will be not 0.001 of the original, but 0.006 of it, and so the drop in intensity will be not 30 dB, but only about 22 dB. On the other hand, if the absorption coefficient were 0.75, and thus the reflection coefficient 0.25 (or the square of what it was in our original example), then the attenuation after each reflection would be as much as it was after each *two* reflections in our original example, and hence the intensity would drop by 60 dB after only 10 reflections, whereas it took

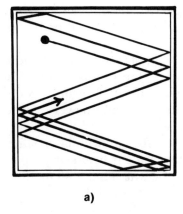

a)

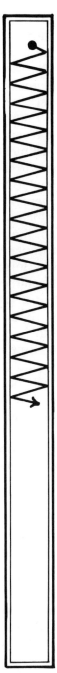

Figure 13.4 Paths of sound waves and their reflections in two different auditoriums with the same areas. We see that in the square one the average path length between reflections is much larger than it is in the narrow and long auditorium.

b)

20 reflections in our original example to attenuate by so many decibels. In other words, the reverberation time would be half of what it was for 0.5.

We see then that if we change the original reflection coefficient to some power of it, we get a reverberation time that is the original reverberation time divided by that power. In the foregoing example, taking the reflection coefficient to the second power divided the reverberation time by a factor of two.

In most books on acoustics, you will find a formula or statement to the effect that the reverberation time is inversely proportional to the absorption coefficient of the wall. That statement is, in general, incorrect, as you can easily see, since if the absorption coefficient is 1.0, all sound energy is absorbed at the first reflection, the sound will reach the observer directly from the source without reflections, and hence the reverberation time is zero. The above formula does not give you this. To be sure, the other extreme case agrees with the statement—namely, if the absorption coefficient is 0.0, then no sound energy is lost on reflection, and the reverberation time is infinite as the sound waves bounce back and forth forever. It turns out that the statement that the reverberation time is inversely proportional to the absorption coefficient is a reasonable approximation if the absorption coefficient is fairly small, that is, much closer to 0 than to 1. From Figure 13.1 you can see that for quite a few materials, and for many frequencies, the absorption coefficient is in fact closer to 0 than to 1, so in those cases it is permissible to use the inverse proportionality with the absorption coefficient since it is then a good approximation to the exact answer which we stated in the previous paragraph.

A Wall of Several Materials

If the hall has various segments of wall surface, each made of a different material, one needs to calculate the amount of sound wave energy loss for each part separately. Let us take as an example the case where half the wall area is made of a material with a reflection coefficient of 0.6 and half of it of a material with a reflection coefficient of 0.4. We can then say that, on the average, the sound wave bounces half the time from that part of the wall that has a coefficient of 0.6 and half the time from that part of the wall that has a coefficient of 0.4. Thus after, say, 10 reflections, we will have a remaining sound intensity that is $(0.6) \times (0.6) \times (0.6) \times (0.6) \times (0.6) \times (0.4) \times (0.4) \times (0.4) \times (0.4) \times (0.4)$ times the intensity before the first reflection. Clearly the remaining intensity, which is 0.0007962 of the original, is somewhere between the 0.006047 of the case when the reflection coefficient of the entire wall is 0.6 and the 0.000105 of the case when the reflection coefficient of the entire wall is 0.4. Clearly there is some intermediate "average" reflection coefficient that produces the same overall attenuation when applied 10 times in succession as the attenuation caused by the succession of five 0.6s and five 0.4s. One might guess that this average coefficient is halfway between 0.6 and 0.4 (that is, 0.5), but that is not true: 0.5 applied 10 times would give 0.0009766. It turns out that this average coefficient is the square root of the product of the two coefficients, or 0.49.

In situations in which we can use the approximation of the reverberation time being inversely proportional to the absorption coefficient, the calculation of the reverberation time that with several different wall segments is simple. One can simply make the reverberation time inversely proportional to the sum, over all wall segments, of the area of that segment multiplied by its absorption coefficient. In our example that would give 0.4 times half the area plus 0.6 times half the area, which is the same as 0.5 times the whole area. You see that the result is only approximate, in that the real average coefficient is not 0.50 as this approximation would have it, but 0.49.

We find the reverberation time of particular interest because it can provide us with much understanding of the way in which combinations of *reflections* produce an overall reverberating sound effect. You can then use the concepts and ideas covered here to approach other acoustic problems, also.

Air Absorption

Sound waves are absorbed not only upon reflection from a wall, but also as they travel through the air of the hall. In halls of normal size, that attenuation is usually small compared with the attenuation by reflection. To illustrate this note that a sound wave with a frequency of 3000 per second, when traveling 200 feet in air at

70°F and a relative humidity of 60 percent, will attenuate only by a factor of about 0.6. The attenuation is some five times larger at a frequency of 10,000, and it also changes significantly with humidity. Since sound waves bounce back and forth, the air absorption is often characterized by a number that gives the attenuation factor by a cubic meter of air in the volume of the hall. Typical amounts in a typical hall are about 10 percent of the attenuation attributable to reflections.

What Is a "Flat" Surface?

We have seen that if we have a plane flat surface, the reflected wave forms the same angle with the surface as the incident wave did. But what is a plane flat surface? Let us assume, for example, that we have an ornate wall with patterns as shown in Figure 13.5. Will this be considered flat by the incident sound wave?

The answer is "yes and no." Just as when we discussed interference of waves, here, also, the basic scale of length is given by the wavelength of the wave. An object that is considerably smaller than the wavelength of the wave will "not be seen" by it. Thus whether the wall in Figure 13.5 will be considered flat by the sound wave depends on the dimension of the ripple. For example, if the ripple is about three feet, then a sound wave with a wavelength smaller than about three feet (or frequency greater than about 300 per second) will find the wall not flat, and hence different parts of the wave will reflect in different directions, depending on the orientation of the part of the wall where they impinge. If, however, the wavelength of the wave is larger than about three feet, or its frequency less than 300 per second, the wave will reflect as if the wall had an orientation "averaged" over the ripples. The figure shows these two cases.

There are many similar situations in connection with other waves when it is obvious that the wave's "eye" has a resolution that is about one wavelength of the wave. For example, we know that ordinary material consists mostly of void, of empty space; since the nuclei are tiny, the electrons, themselves tiny, circulate around nuclei very far away compared with their own sizes, with empty space in between, and even atoms consisting of these nuclei and electrons are located at distances that can be considerable

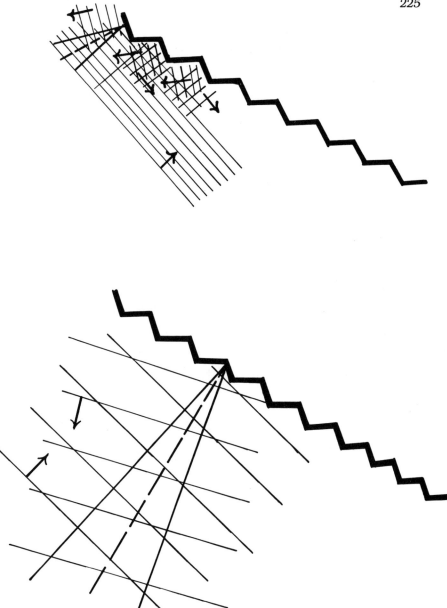

Figure 13.5 A sound wave "sees" irregularities in the surface on which it reflects, if the wavelength of the sound is the same as or smaller than the size of the irregularity. In the upper illustration, the wavelength is small, and therefore the reflection is determined by the orientation of each segment of the wall surface. Hence some of the wave reflects to the left, and some to the right. In the lower illustration, the wavelength is large, and hence the wave reflects from the average of the wrinkled wall, as if the wall were a smooth plane and the wrinkles would not be there at all.

compared with their own sizes, again with empty space in be-tween. Yet when we look at material, we see it as solid. This is so because the wavelength of the visible light is very much larger than the spacing between nuclei, electrons, or atoms, and hence to that light wave the intermittent lattice of these particles ap-pears continuous.

To take a different example, you may have noticed that often the "dishes" used to receive television or other signals from satellites are not made of a solid sheet of metal but of a metal net. It may seem that the waves falling on the netlike surface would go through it because of the holes in the net. They do not, because the spacing of the net is designed to be smaller than the wavelength of the signal waves the dish is supposed to receive, and hence to that wave the dish is solid, even though to our eyes, which use visible light (with a very much smaller wavelength than that of the televi-sion signals), the dish is mostly holes.

To return to sound, we see that if we have a rippled surface on the wall of a hall, the reflections of sound waves at various frequen-cies will be quite different. The high-frequency sound will be dis-persed in all directions, while the low-frequency sound will be reflected in a definite direction. This represents a tool with which an architect can regulate the distribution of sound in the hall. Since other influences on the sound waves are also frequency dependent (e.g., the absorption of sound in air), this frequency dependence of the direction of reflection can be used to counteract frequency distortions resulting from those other effects. Or we might want to enhance the "brilliance" (i.e., high-frequency con-tent) of the sound by utilizing the different frequency responses of the wall to reflections.

Initial Time Delay

The first sound to reach the listener from the source is the one that propagates directly, in a straight line (unless some obstacle prevents this). The reflected waves will reach in close succession, but after an initial time delay that represents the path difference between the direct straight line and the shortest reflected path (see Figure 13.6). In the example in that figure, this initial time delay can be calculated by measuring the direct path SL (which is 72 feet), and then measuring several candidate path lengths to find the second shortest. We find that SAL is 87 feet, SBL is 95 feet, and SCL is 101 feet. The difference between 87 feet and 72 feet is 15 feet, which represents (15 feet)/(1000 ft/s) = 0.015 second delay. The delay time in this example is of the typical magnitude for time delays found in actual concert halls. We also see from this example that the subsequent reflected waves arrive closer to each other, so that the initial time delay is indeed the largest time difference among sound signals.

Applications

We are now ready to see how the knowledge we acquired about the propagation and reflections of waves can be used in the practical aspects of the music environment.

In an outdoor environment, the musicians often have a shell behind them to reflect the sound waves that otherwise would go off in a direction in which there were no listeners. Shells are used even in indoor settings, to get the sound out of the stage area and into the audience.

Some performances are held in sheds with a roof but no side walls. Here the roof can be used to reflect sound, but horizontally sound can propagate in all directions and hence there will be a decrease in sound intensity, proportional to the distance from the stage, as we saw when we discussed scaling and the propagation of waves in Chapter 4. Sheds are also ineffective in screening out outside noise.

Most of the attention, however, is directed toward the design of all-enclosed halls. We can now return to our list of requirements and see how one can satisfy them.

The hall can be shielded from external noise by careful insulation of the walls. We saw that as a sound wave (this time from the outside) impinges on a surface, some of it refracts and penetrates into the wall. In the wall it continues to propagate, but it gets attenuated while doing so because of absorption by the material that makes up the wall. We saw that in air this absorption by the medium itself is small even over a path length of 200 feet, but if the

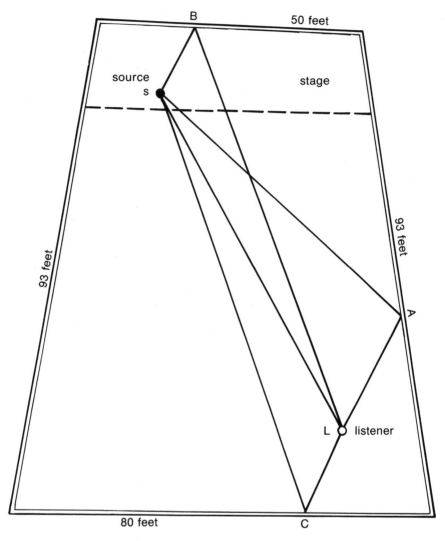

Figure 13.6 Many paths of the sound wave, after one reflection, reach the listener. As they are of different length, the indirect sound does not reach the listener as a sharp impulse at a definite time, but is drawn out over an extended period of time.

wall is made of specially designed material, this absorption can be made enormous even over a short distance. A good test of this technology is found in the construction of the Kennedy Center in Washington, D.C. which is under the flight path of planes using the not very distant National Airport. Yet performances in that building are completely shielded from the noise of the aircraft. Fluffy, fibrous material is particularly suitable for such insulation.

To make the sound of sufficiently high volume for everyone in the auditorium basically requires two factors. First, the auditorium needs to be sufficiently small for the particular sound source. To present a solo recorder recital in the Shrine Auditorium of Los Angeles, which holds 6000 or more persons, would be folly. Different sized halls thus are used for different types of music: halls that hold 500–1000 for solo recitals and chamber music, and larger ones for orchestral concerts, musicals, and opera.

The other factor involved is that there must be no "dead spots" in the hall, that is, locations where the sound is anomalously soft. Such dead spots can result from the obstruction of the direct sound or of some few-times-reflected sounds, or from destructive interference between various reflected sound waves. Consider Figure 13.7 in which two reflected waves, with a difference in path length corresponding to half a wavelength, are shown to interfere destructively at a location. In general, of course, there are reflected waves coming from various directions and at various times, and so one would expect that these waves interfere with each other partly constructively and partly destructively, and on the average about the same way regardless of the location of the listener. There may, however, be particular locations where this is not true. To avoid them, one should make sure that each location is reached by a large multitude of reflected waves from all directions, so that one can count on this uniform averaging out.

It is assumed here that no artificial means are used to enhance the volume of the sound. Often such electronic amplification is used, especially for popular (as distinct from classical) music. That may solve the sound-intensity problem, but can create problems of distortion and balance (discussed in the next chapter).

The next feature on our list is clarity. Here we must realize that the terminology used in describing the requirements of halls is to a large extent that of music lovers and critics, and can be characterized as subjective and psychological rather than scientific. The

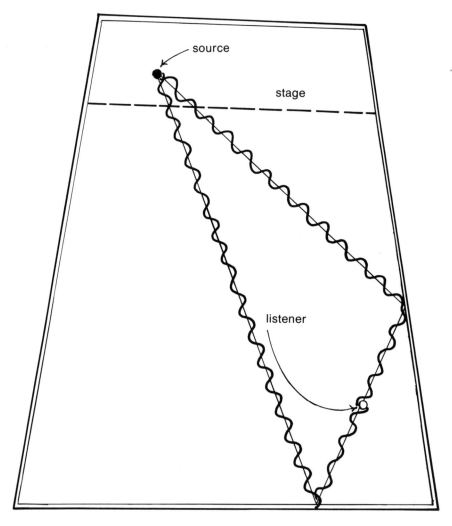

Figure 13.7 Two paths of a sound wave that arrive at the listener out of phase. Such interferences can create dead spots in auditoriums.

"translation" from one "language" to the other has been accomplished through psychoacoustic experiments in which certain physical environments are created and listeners are asked to describe them in terms of their own subjective vocabulary.

We know in this way that what we mean by clarity and liveliness pertains to the length of the reverberation time. If this time is long, consecutive sound impressions merge into each other and become garbled. On the other hand, too short a reverberation time makes the melodic line of the music disjointed and abrupt. Furthermore, a short reverberation time (and, in general, a very soft reverberant sound compared with the direct sound) will destroy the sound's liveliness. We can see that our requirements are in part mutually contradictory, calling for opposite properties for, in this case, the reverberation time.

Being reached by sound from all directions is also important for the creation of the effect of "being enveloped in sound." The quadriphonic recording is designed to accomplish this in one's home. For this purpose it is important not to drape the back walls of the hall with highly absorbing material, because if we preclude reflection from the back wall, only sound from the front will reach the listener.

We also mentioned, as a requirement, that the conditions we specify be equally valid for both low and high frequencies. In the listener's terminology, the lack of low frequency will be described as a lack of warmth, and the lack of high frequency as a lack of brilliance.

We mentioned the requirement of spatial perception. In a large ensemble of musicians, the listener wants to be able to differentiate what originates in various parts of the stage. With direct sound only, this would not be difficult, as we explained in Chapter 5. With reflected sound, however, it is much more complex.

Another requirement was the absence of an echo. An echo is a reflected sound of substantial volume that occurs separately, and with a noticeable time delay after the direct sound.

Many of these considerations depend on the reverberation patterns and times. What is the optimal value for the latter? There is no single answer, since it depends on the use to which the hall is being put. For orchestral music it is around 1.5–2.5 seconds; for chamber music and small ensembles, 1.0–1.5 seconds; and for speech and drama, between 0.5–1.0 second. This variation in optimal reverberation times is one of the main challenges in the design of a multipurpose auditorium, of the type found in many medium-size cities. Certain special environments have (and are expected to have) special reverberation times. We are used to old cathedrals having a cavernous quality, which means a reverberation time of as long as four seconds. Modern churches, while still quite reverberant, have a reverberation time half of that.

The attenuation of the sound during the reverberation may not be according to what we had when considering a uniform hall with the same wall covering throughout. In that case the attenuation, as a function of time, was a decreasing exponential, if we plotted intensity, or a straight-line decrease, if we plotted decibels (see Figure 13.8). In reality, however, because of the uneven patterns of reflection, complicated interference effects, and different wall surfaces, the attenuation may be different, as also shown in Figure 13.8. These deviating patterns are undesirable from the standpoint of the listener.

What means are at our disposal to influence reverberation time? There are many: the size and shape of auditoriums, the type of wall coverings, and movable "clouds" (i.e., suspended movable reflecting surfaces), as well as artificial supplementation of sound through electronic means. There are also factors beyond our control, such as the type of clothing worn by members of the audience (which can significantly influence the absorption coefficient of the floor of the hall).

Is Science Successful in Designing Halls?

In view of all the above, can we design an ideal auditorium on the basis of scientific considerations alone? Certainly not. As we mentioned in our earlier discussion of science and technology, what science can provide in attaining this technological objective of a good auditorium is some broad outlines and conditions. Beyond that, practical experience and experimentation through trial and error are needed to zero in, within the confines provided by science, on the ideal hall. It is noteworthy that even today, some of the most successful auditoriums, such as Symphony Hall in Boston, were built more than 80 years ago when the science of sound was not yet sufficiently extended to practical considerations to be of much use in this respect.

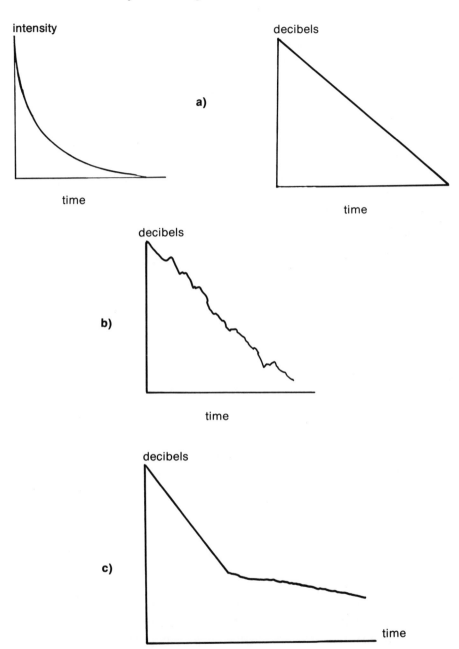

Figure 13.8 The attenuation of sound intensity with time. The pattern shown in **(a)** is the "ideal" exponential decrease one would expect from calculations under simplified circumstances. Patterns **(b)** and **(c)** shows different patterns that may occur in practice.

Some other old halls, however, are terrible, and it is safe to say that today the construction of halls can be expected, on the average, to be more successful than it was a century ago, provided that the designer, using scientific principles, and the actual constructor work closely together. An example of where this was supposedly not done, is Philharmonic Hall in New York, which had to be completely reconstructed inside, at a cost of many millions of dollars, when the original version of the hall was deemed very unsatisfactory acoustically. One unfortunate aspect of the situation is, however, that relatively few auditoriums are constructed, and no two are identical, so that learning from experimentation with new halls is not easy. Few communities would tolerate a novel but mediocre auditorium because the knowledge that was gained from the failed experiment incorporated in the auditorium would benefit, in the long run, the science and art of auditorium design.

Summary

In this chapter we discussed the requirements for a favorable environment for listening to music in a closed space and the ways to satisfy those requirements. The most important aspect of this discussion concerns the way waves reflect from the walls of the hall. We used the absorption (or reflection) coefficients to describe the fraction of the wave energy that does not (or does) reflect after the wave hits the wall. The reverberation time was defined as the time it takes for the sound wave, reflected many times, to drop 60 dB below its original intensity. The reverberation time depends on the dimensions of the hall and on the absorption coefficient of the walls. We deduced that the reverberation time, in particular, is proportional to the volume of the hall and inversely proportional to the surface area of the wall. The dependence on the absorption (or reflection) coefficients is, in general, more complicated, but for relatively small absorption coefficients, we can say approximately that the reverberation time is inversely proportional to the absorption coefficient. In this approximation for a hall with several wall areas with different absorption coefficients, we have to multiply each coefficient by the area of the wall having that coefficient, and then add up the various products. Some sound energy is also absorbed by the air in the hall, especially at higher frequencies. Reflection from a flat wall is at the same angle as incidence (but on the other side of the perpendicular to the wall). If irregularities in the flatness of the wall are much smaller than the length of the wave, the wave will not "see" them and average over their directions.

In judging sound effects, the initial time delay plays an important role. This is the difference in time between the arrival of the direct sound from the source and the arrival of the first reflected sound wave. The listener's requirements expressed in terms of a figurative language (e.g., "warmth," "brilliance") can be translated into physical conditions through psychoacoustic experiments. In this way one can take technological steps to satisfy these requirements. The hall needs to be insulated from outside noise. To ensure sufficiently high-intensity levels for all listeners in the hall, the hall's size has to be matched by the volume of sound put out by the source, and dead spots in the hall caused by obstructions and destructive interference must be eliminated. The reverberation time must be sufficiently long to assure warmth of the sound, but sufficiently short to avoid individual notes to blur into each other.

Sound needs to reach the listener from all directions in order to give an impression of being enveloped in sound. The balance between low and high frequencies in the music can be assured by controlling those aspects of sound propagation and reflection that are frequency dependent. Other requirements include spatial perception of sound, an absence of an echo, and a proper time history of the reverberating sound. Ideal reverberation times depend on the particular use of the hall, and are longer for orchestral music, less for chamber music and solo recitals, and even less for speech and drama. All this scientific foundation helps only to find the *approximate* specifications of a good hall, beyond which trial-and-error technology, guided by experience and experimentation, needs to be applied to achieve a good design.

Chapter Fourteen

The Transmission and Storage of Sound

The Objectives

T HERE ARE TWO major deficiencies of sound from the point of view of making a musical performance enjoyable to a large number of people: it fades fast in relation to both distance and time. We have seen that the sound intensity drops very fast as we recede from the location of the sound source, and that the energy of the sound wave is quickly converted into heat, thus erasing the record of the music that was performed. As a result of this, in former times the enjoyment of superb music and musical performances was limited to a tiny fraction of the population. The total number of people during the 19th century who heard anything written by Beethoven performed by professional musicians could not have been more than 0.1 percent of the number of people who lived in that century, considering that concerts were attended mainly by members the urban upper middle class, and that the same people tended to go to concerts over and over again. Such concerts were the only opportunity then to hear such music. The situation was even more restrictive in earlier centuries.

Technology has changed all that. Today a performance given by the Metropolitan Opera is potentially available to 240 million people, and actually listened to by tens of millions. A similar number has the opportunity to listen, in their own homes, and whenever they wish, to at least a dozen compositions by Beethoven, performed by the world's greatest artists some of them long dead. The same holds for the products and performers of so-called popular music. The enjoyment of music today in any of the technologically advanced countries is limited, in the main, only by one's will to take advantage of the opportunity.

In this chapter we discuss the basic elements of the technology that allows this unlimited dispersion of music in space and time.

Figure 14.1 shows the overall principles. Whether we are concerned with transmission of music simultaneously with the performance but over large distances, or with making a permanent record of the music that can be recreated at a later time, these principles are, for the most part, similar, and only some ingredients in the chain of the process differ for the two cases. For this reason the discussion of the transmission of music can also be utilized for much of the discussion of the storage of music.

Why Not Do It Directly?

Before we start to discuss the transmission of sound over large distances, it is instructive to explore why we cannot do it directly, without the conversion processes described in Figure 14.1. After all, we know instances where sound travels quite far. The sound of thunder can be heard for many miles, and I heard the sound of the eruption of Mount St. Helens, even though at the time I was hiking in the mountains of Oregon, some 120 miles away.

The first problem is that the sound we want to transmit occurs at much lower intensity levels than thunder or a volcanic eruption, and we have no good way to amplify such a low-intensity sound to very-high-intensity levels in a direct way, that is, without utilizing the type of conversion processes shown in Figure 14.1. Second, even if we did, such an intense sound source would be impossible to bear at close distances. A clap of thunder at a distance of a quarter of a mile is deafening, and even that sound is transmitted only over, perhaps, 10–20 miles. Third, sound waves attenuate too much. Consider the following example. Let us take a total sound energy per second equivalent to the energy per second emitted by a fairly strong AM radio station, which is 100,000 watts. We will now

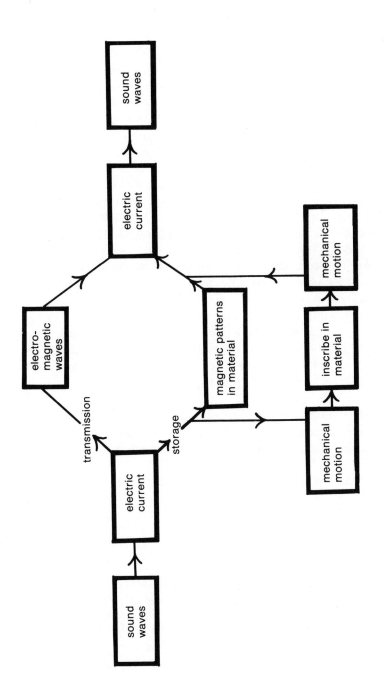

Figure 14.1 Schematic diagram of the transmission and storage of sound. Note that the first and last steps are common to all methods.

calculate what we would hear of this sound, say, 100 miles away. First let us take the attenuation attributable to the distance alone. The surface area of a sphere with a radius of 100 miles is about 12 $\times (100)^2 = 1.2 \times 10^5$ square miles, or about 3×10^{12} square feet. The sound-gathering area of our ear is four square inches at the most, or about 1/30 square foot. So our ear would, at a distance of 100 miles, receive only 10^{-14} times the original sound energy, if the loss were due only to the distance alone. But then there is also absorption in the air, which, as we said in the last chapter, amounts, at a frequency of 3000 per second, to something like a factor of 0.6 every 200 feet. Since there are about 25 distances of 200 feet in a mile, the attenuation per mile due to absorption in air is 0.6^{25}, or about 2×10^{-6}. For 100 miles this gives a factor of 2×10^{-600}, such a small number that the sound energy reaching us after 100 miles of travel through the air would be completely negligible.

In view of this incredibly tiny number, one might wonder how I heard Mount St. Helens 120 miles away. What I heard was a very-low-frequency rumble, and you recall that the absorption in air depends very much on the frequency of the sound wave. The factor of 0.6 we used above was for a frequency of 3000 per second. For a frequency of 10,000 per second, it is 0.12. Going in the other direction, the factor for, say, a frequency of 100 per second would be only about 0.992, and with that factor the attenuation is only $0.992^{26} = 0.81$ per mile, or $0.81^{100} = 1.7 \times 10^{-9}$ or 87 dB over 100 miles, which still allows one to hear a strong sound at that distance. This fact is utilized by Africans who can communicate, using a very low-frequency bongo drum, over a distance of quite a few miles.

In any case, however, we can safely conclude that a direct transmission of sound over long distances at the frequencies at which most of our speech and music take place is not feasible.

Changing Sound Into Electric Current and Vice Versa

We shall discuss each step in Figure 14.1, starting with the conversion of sound waves into electric current. The key element in this process is the phenomenon that, if we move a piece of wire when it is close to a magnet, an electric current will start flowing in it. This current is larger if the magnet is stronger and the motion is faster. When the motion stops, the current also subsides.

If we know this, it is not difficult to figure out a way to convert sound waves into electric current. We let sound waves impinge on some membrane that conducts electricity, as the wire does, and place this membrane close to a magnet. The membrane will vibrate as the sound waves impact on it, and correspondingly a fluctuating current will appear in it that can be conducted away in a wire attached to the membrane.

The contraption we just designed can also be used backwards; if we lead electric current through a wire or some other current-conducting object that is close to a magnet, the object will move. If the current fluctuates in time as a sound wave would, the conducting object will do likewise, and by its motion will generate sound waves. Thus electric current fluctuations can be reconverted into sound waves.

The technological name for this device is "microphone," if it turns sound into current, and "loudspeaker," if it works in reverse. In actual practice the two look quite different, but the principles are the same.

This conversion process can be called electromagnetic since both electric current and a magnet are parts of it. There is also a somewhat different, purely electric way of converting sound to current or vice versa; it is called "electrostatic." Here we place two conducting plates close to each other (see Figure 14.2), and place electric charges of opposite signs on the two plates. This pair of plates can thus "hold" charges, because charges on the two plates attract each other and so are kept from flowing away through wires attached to the plates. The amount of charge such a pair of plates can hold, however, depends on the distance between the two plates, since the forces between the charges on the two plates also vary with the distance. If, therefore, we let sound waves move one of the plates, thereby varying the distance between the two plates, the amount of charge that can be held on the pair of plates will change, and so some current will either depart from or arrive at the plates. Moving charge is current, and hence we have managed to convert sound waves to current. This method can also be used in reverse, to convert current into sound waves.

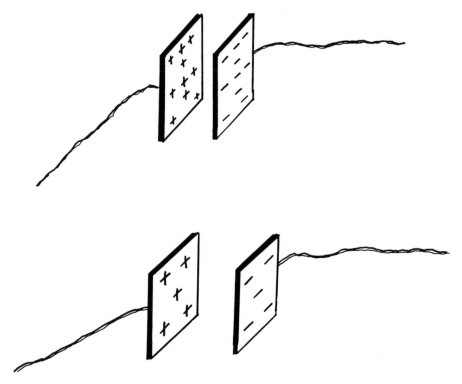

Figure 14.2 The principle of an electrostatic microphone or loudspeaker. The amount of charge that accumulates on the plates depends on the distance between the plates. Therefore, if the distance changes due to the force of sound waves on the plates, some charge must come to the plates or flow from them, thus creating a current.

Converting Electric Current Into Electromagnetic Waves

Electric current consists of moving electric charges. As these charges move, they also radiate away some of their energy, just as an ember radiates away some of its heat energy. You can notice this heat radiation by placing your hand close to but not in contact with the ember, and feeling the heat coming from it. We do not similarly feel the electric current radiating away its energy, but only because the heat radiation from the ember and the radiation from the current are on different frequencies, and our bodily sensors are sensitive to one and not the other.

We can think of this radiation of electric energy away from the current as a wave pouring out of the current. A wave of what? For our purposes we can just say that electromagnetic waves are generated, which are carriers of energy away from the charges. Their nature is the same as that of any other electromagnetic wave, such as visible light, x-rays, and ultraviolet light. We discussed this briefly in Chapter 3. As we saw there, the essential and strange property of such electromagnetic waves is that they do not need a material medium in which to propagate. On the contrary, the emptier space is, the easier (and faster) they propagate. We know this, of course, since we can see the sun in spite of the almost completely empty space between it and the earth through which its light must propagate to reach us.

The conversion of current to such electromagnetic radiation is accomplished rather efficiently in wire networks called antennas, in which charges move back and forth, and while doing this constantly radiate their energy. Similarly, antennas are also used for the reverse process: charged particles inside conducting material will start moving and forming a current when electromagnetic waves fall on the antenna and impart their energies to these electric charges. Thus the process works equally well both ways.

Propagation of Electromagnetic Waves

The only link in the upper chain of processes in Figure 14.1 that we have not yet discussed is the propagation of the electromagnetic waves from the transmitting antenna to the receiving antenna.

Attenuation that is due to distance alone, of course, affects electromagnetic waves just as it does sound waves. We saw, however, in our earlier discussion of why the direct method of sound transmission does not work, that the attenuating factor due to distance alone is not very large. For sound waves the really incapacitating effect was the attenuation caused by absorption in air. This absorption is very much less in the case of electromagnetic waves, and in fact some electromagnetic waves (with appropriately chosen frequencies) can travel thousands of miles in air without being severely absorbed.

The next problem we have to be concerned with is obstructions in the path between the emitter and the receiver. Here we must recall that the "eyes" of waves can only see obstacles that are the same size as or larger than the wavelength of the wave. What are these wavelengths in the case of the electromagnetic waves we use to transmit music? As you will see, these wavelengths have nothing to do with the wavelengths of the sound waves that are to be transmitted. What matters is the wavelength of the "carrier wave," (see below), which can be chosen at our convenience to satisfy the various requirements of the technology used in this process.

As it turns out, we have been using a range of wavelengths for this purpose. The "100" marking on an AM radio dial, for example, corresponds to a wavelength of about 1000 feet. In contrast, the 100 on an FM dial corresponds to waves with a wavelength of about 10 feet. As a result obstacles such as a house or two, or even a small hill, do not impede AM waves very much, but for FM reception even relatively small objects are bothersome and one needs a truly direct "line of sight" to the emitting antenna.

There is also another reason why AM radio waves reach a larger fraction of receiving antennas within their range than FM waves do. Those electromagnetic radio waves that propagate upward into the earth's atmosphere normally would eventually leave the earth, just as light waves from the surface of the earth also leave the earth (which is why the earth is visible from the moon). It turns out, however, that there is, around the earth, at an altitude of about 80 miles, a layer of charged particles, and this layer causes the electromagnetic AM radio waves to reflect down again (see Figure 14.3). Incidentally, this type of charged particle layer is also responsible for the temporary blackout in communications that reentering spacecrafts experience. In any case, because of the reflections of the AM waves from this layer, we can receive the waves emitted by AM radio stations at quite large distances at which, due to the earth's curvature, line-of-sight reception would be impossible. The layer is formed as a result of radiation from the sun, and for that reason, at night the layer is at higher altitudes than during the day. Reflection from higher altitudes means a greater range, which is why certain distant stations, not available in the daytime, can be received at night.

This reflecting property of the layer of charged particles acts only on electromagnetic waves in a certain range of wavelengths. The AM radio waves fall into this range, but the FM waves do not.

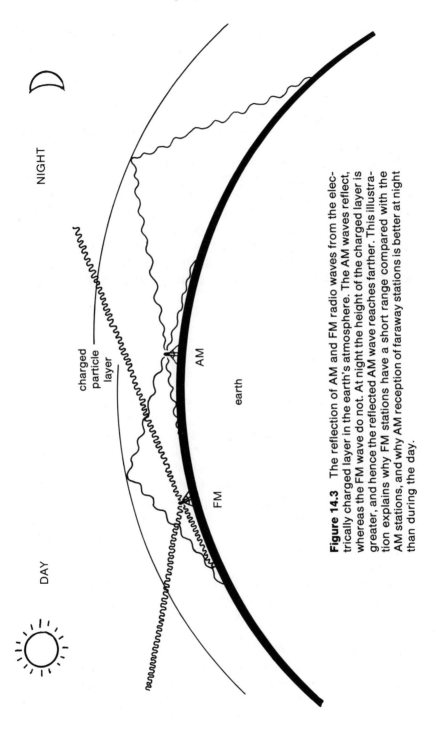

Figure 14.3 The reflection of AM and FM radio waves from the electrically charged layer in the earth's atmosphere. The AM waves reflect, whereas the FM wave do not. At night the height of the charged layer is greater, and hence the reflected AM wave reaches farther. This illustration explains why FM stations have a short range compared with the AM stations, and why AM reception of faraway stations is better at night than during the day.

As a result FM radio waves do not bounce back and hence FM transmission can be received only by direct line of sight (or nearly so).

A final and practical reason why FM stations have a much shorter range than AM stations is that FM stations generally emit much less energy than AM stations. Indeed, the FM technology has been used for small, local, and often special-interest radio stations, thus lending a much greater variety and range of style to radio entertainment than was possible before the FM era.

On the other hand, there are situations when the FM waves have the upper hand in propagating. You may have noticed that inside large buildings with steel frames, AM reception is difficult and FM reception is better. This is so because to the AM waves, with their large wavelengths, the steel beam skeleton of the building, with a spacing of 10–20 feet, is impenetrable, while for FM waves, which have much smaller wavelengths, penetration is possible.

Carrier Waves and Their Modulation

If all radio stations broadcast directly on the frequencies of the music they wanted to transmit, there would be chaos and a confusing overlap of the stations, among which we could not choose. For this reason, as well as for others involving the technology of radio equipment, each station is assigned a *carrier wave* of a certain frequency at which they broadcast. The carrier wave itself is a regular sine wave with a given frequency, which, from the point of view of what we are used to in connection with sound waves, is very high. An AM dial ranges in frequency from 550,000 to 1,600,000 per second, whereas an FM dial comprises carrier waves with frequencies from 93 million to 106 million per second.

The sound wave, turned into electric current fluctuations, is then *superimposed* on this carrier wave, thus changing its shape from a regular uniform sine wave to something else that is characteristic of the waveshape of the music we want to broadcast. There are two ways to do this.

The AM (or amplitude-modulating) emitter changes the *amplitude* of the carrier wave in accordance with the amplitude of the sound wave that has been converted into current fluctuations. Figure 14.4 shows this. Note that the scale on that figure is wrong, and was used only to make the drawing possible. In reality a standard A with a frequency of 440 per second, when superimposed on the carrier wave with a frequency of 1,000,000, would have a modulating peak only at every 2273th peak of the original carrier wave, and not at every 10th, as shown in the figure.

The other method of superimposing the music wave onto the carrier wave is to modify the *frequency* of the carrier wave according to the amplitude of the music wave. This is shown in Figure 14.5. Again, the proportions are incorrect and were chosen only to enable the effect to be seen, and in fact in this case for two reasons: first, the fractional variation of the frequency of the carrier wave due to modulation is, in actual practice, less than one-fifth of 1 percent; and second, "peaks" in this modulation would appear at every 250,000th peak of the carrier wave and not at every 10th as in the figure.

Avoiding Noise

Even without radio transmitters, our environment would be far from free of electromagnetic waves in the general frequency range of the radio waves. Other man-made electromagnetic waves include radiation from electric motors of all sorts, in homes, in vehicles, in factories, and so on. Nature also contributes its share to the noise. Atmospheric disturbances such as thunderstorms, or even just static electricity in the air, result in such electromagnetic waves. All this causes the "static" you hear when you listen to radio stations, especially if they are far away.

It turns out that such electromagnetic noises are more likely to occur in the frequency range corresponding to the AM radio band than in the range of the FM band. For this reason FM reception is usually much freer of static than that from AM stations.

Amplification

Even if electromagnetic waves do not attenuate very much by absorption in the air, the signal reaching the receiving antenna is

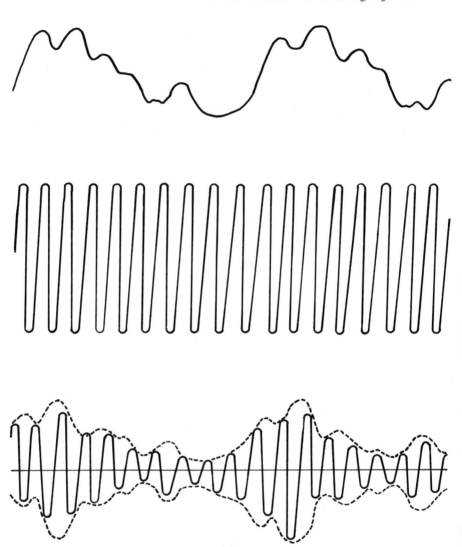

Figure 14.4 The principle of amplitude modulation (AM). At the top is the sound wave we want to transmit. the second line shows the unmodulated carrier wave. The third line is the carrier wave with its amplitude shaped according to the amplitude of the sound wave. In reality the frequency of the sound wave is not about 10 times smaller than the frequency of the carrier wave (as shown here), but about 1000 times or more.

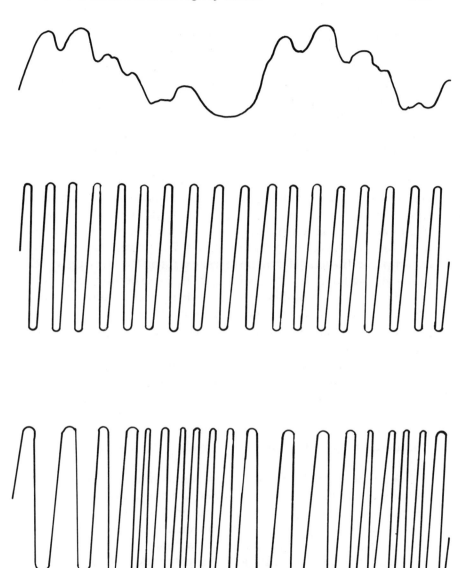

Figure 14.5 The principle of frequency modulation (FM). Just as in Figure 14.4, the top line is the sound wave, the second line is the unmodulated carrier wave, and the third line is the modulated carrier wave, the frequency of which was changed according to the amplitude of the sound wave. In reality the frequency of the sound wave is not about 10 times smaller than the frequency of the carrier wave (as shown here), but about 100,000 times or more.

likely to be very weak compared with the emitted wave, because of the distance effect. Hence the tiny signal that is picked up by the antenna, and the subsequent tiny currents that are created in the antenna, contain too little energy to produce an audible sound wave. Thus the current fluctuations first must be amplified.

Such amplifications also play an important role in the additional links related to the storing of music, which we discuss presently, and so it is doubly important for us to say something about amplification.

The term "amplification" is used in science and technology to denote a situation in which a small effect is made large by using a second phenomenon that is very sensitively influenced by the first phenomenon in which the small effect takes place. There are innumerable examples of such situations. The power steering on a car amplifies the small force the driver exerts on the steering wheel into a strong force that steers the car. Those experimenting with artificial weather modification, and with cloud seeding in particular, try to produce large changes in cloud precipitation by injecting small changes into the physical conditions in the clouds. The famous bass aria "La Calumnia" in Rossini's *The Barber of Seville* describes how a few slanderous remarks about somebody, casually dropped into conversations with people in a community, become amplified into an avalanche of ill feelings toward that person. In fact, snow avalanche itself is also an example of amplification: a small dislodged rock or clump of snow, in falling, accumulates more snow and rock, ending in a huge mass of moving material.

In each of these examples, a primary phenomenon influences a secondary phenomenon. The former are the movement of the steering wheel, the injection of crystals into the clouds, the injection of false information into the conversational stream, and the dislodging of a rock respectively. The secondary phenomena are the turning of the front wheels, the precipitation of water in the clouds, the communicative flow in a human community, and the tendency of things to roll downhill and for flakes of snow to stick together respectively. The first of these examples is the closest to what we are after, since in it continuous control over the second phenomenon can be exercised; the other examples tend to be "yes-or-no" propositions, which, once triggered, tend to get out of hand.

The particular kind of amplification of concern to us now is that of electric current. Here we note that the amount of flow of charged particles, in either the old-fashioned vacuum tube or the modern transistor, depends very sensitively on the value of the voltage of a regulating obstacle that we place between the device's emitter and collector. Thus small currents from the primary signal can be used to regulate the voltage on the obstacle, and the fluctuations of this small current will then produce a much larger fluctuation in the flow of charged particle, which we will then use as the secondary, amplified current.

A very important requirement of amplifiers is that they be distortion-free, that is, we want the shape of the secondary signal to be the same as the shape of the primary signal. To describe this situation, we use a response curve (see Figure 14.6), which tells us the amount of change in the secondary current as a result of a given change in the primary current. Depending on the level of the primary current, the ratio of these two changes may be different. For a distortion-free situation, this ratio should be independent of the level of the primary current; that is, the response curve should be a straight line.

Methods of Storing Music

We now turn to the lower two chains of the process described in Figure 14.1, namely, the storage of music. As we said earlier, the situation here differs from the transmission of sound only in the few steps in the middle of the chain. So we will now concentrate on these.

There are two commonly used ways to store music: on phonograph records and on tape. In both cases we start our discussion with the sound wave already transformed into electric current fluctuations.

In the case of the phonograph record, the electric current fluctuations are used to induce mechanical vibrations in a fine needle, which then scratches the surface of a disk, leaving an imprint of these vibrations. If we want to recreate the music, we simply reverse the process. We run a fine needle over the scratched sur-

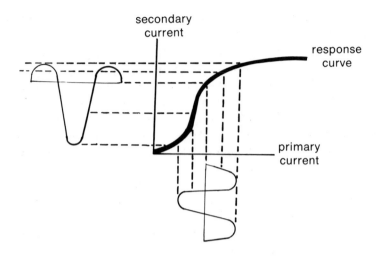

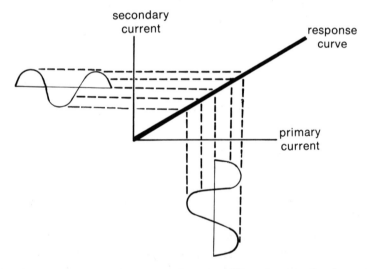

Figure 14.6 The response curve of an amplifier changes the incoming primary wave into the outgoing secondary wave. The primary wave is shown below the horizontal axis. Each point on it is projected (by the vertical dotted lines) onto the response curve and then projected (by the horizontal dotted lines) into the secondary current. Different response curves are shown. The bottom one is a straight line, and so the shape of the primary wave is the same as that of the secondary wave. The top one is a curved line, and hence the secondary shape is a distorted one compared with the primary shape. For a detailed example, see Exercise 14.1.

face, thus inducing vibrations in the needle, which in turn will be converted into electric current fluctuations. The principle of the conversion of mechanical motion to electric current fluctuation, or vice versa, was discussed earlier in this chapter, with regard to the conversion of sound waves into electric current fluctuations.

The needle must be fine so that it can follow the rapid changes in time that take place during vibrations. We know that for a bright overtone structure, we need frequencies up to 10,000 per second, or even somewhat beyond that, and a bulky needle would be unable to respond to such rapid vibrations.

To eliminate this mechanical element in the recreation of music from the record, the most recent recording techniques use a laser light beam to "read" the inscriptions on the record. Since light has, for all practical purposes, no mass, it can follow the fast vibrations very easily. This is the method used to make compact disks.

In the case of storing music on tape, the inscription is done not by causing controlled mechanical damage to a plate, but by rearranging the magnetic patterns in a magnetic material deposited on a tape. One can imagine this magnetic material as consisting of billions of tiny magnets, which, when held close to a magnet, can change the direction of their orientations. Initially these tiny magnets are randomly oriented in all directions. When the tape passes a magnet, the strength of which is controlled by the current that resulted from the conversion of sound waves into electric current, the tiny magnets will partially coordinate their orientation directions, with the degree of coordination dependent on the strength of the outside magnet. The replaying process is the same in reverse. The varying magnetization on the tape produces a varying current in the tape head, which is then converted into sound waves.

Since the mass of each tiny magnet is extremely small, it can respond very easily to very rapid variations in the outside magnetic strength, thus providing an advantage over the phonograph needle. Tapes also keep better without damage, are sturdier in use, and occupy less space. On the other hand, they can be damaged at elevated temperatures, because the heat energy "rattles" the tiny magnets and destroys their coordinated orientations. Also, tapes must be kept away from magnets, of course, since proximity to outside magnetic influence will also disorient the lining up of the tiny magnets in the tape's magnetic material.

Noise Reduction

The magnetic material of the tape, however carefully manufactured, always has some unevenness and inhomogeneities that show up as a hissing background noise when the tape is used for recording. Since the scale of these inhomogeneities on the tape is very small, and hence the variation is rapid as the tape passes the tape head, the resulting noise is mainly at high frequencies. This noise is at low intensities, and as long as the intensity of the music on the tape is considerably higher than that of the noise, the latter is not bothersome. One would therefore think of recording all of the music at very-high-intensity levels, that is, one would amplify the electric current considerably when feeding it into the tape head. Unfortunately that idea fails for another reason. In most musical pieces, there is a considerable variation in intensity levels, from very soft to overwhelmingly loud. If, therefore, we record even the softest passages at a considerable intensity, we will compress the intensity range in the recording, since the tape itself can handle only a certain range of intensities, and we throw away half of this range if we record even the softest passages at the middle of that range.

The solution to this problem is to record the soft passages at a higher intensity and the loudest passages at a lower intensity, and upon replaying, suppress the intensity level of the soft notes while leaving the level of the loud notes the same—in effect, reexpanding the intensity range to what it was in the original music. In suppressing the intensity level of the soft notes, we simultaneously suppress the noise level also, and so the "relative edge" of the music over noise is maintained, which was our original purpose. This is shown schematically in Figure 14.7. Among the various commercial systems utilizing such noise-reduction principles, the most commonly used is sold under the trade name Dolby.

Summary

Since sound fades fast with distance and time, we need methods of transmitting sound over large distances and of storing

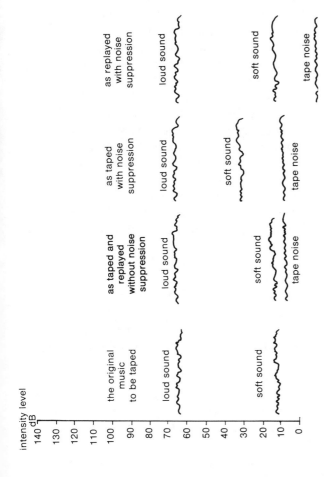

Figure 14.7 The principle of noise suppression in sound recording. The first column on the left shows the sound levels in the original music to be taped. The second column shows the sound levels when such a recording is replayed. The third column shows the sound levels during the recording process with noise suppression. The soft sound is artificially amplified to be significantly higher than the tape noise level. The fourth column shows the sound levels during replaying, when all soft sound levels (including that of the tape noise) are lowered again. We see, by comparing the second and fourth columns, that the difference between soft sound and tape noise is much larger in the noise-suppression method than in the usual method of recording.

music for long periods of time. To transmit music directly over, say, 100 miles is impossible, since even if enormously strong sound emitters were set up, all but the lowest frequencies would be absorbed by the air over such distances.

The process for both transmission and storing involves converting sound waves into electric current fluctuations and then reconverting them into sound waves. This conversion process is accomplished by utilizing the fact that motion of a wire or other conducting object near a magnet will cause the current to flow. Another method is to use a pair of conducting plates with electric charges on them, and make use of the currents that are created by electric charges flowing toward or away from the plates as the distance between the two plates is changed.

Electric currents radiate some of their energy away into electromagnetic waves that can travel even in empty space. The absorption of these electromagnetic waves in air is very little compared with sound waves. The AM-type radio waves have wavelengths of the order of 1000 feet, whereas those of the FM type are about 10 feet. These sizes determine the kind of obstacles that matter when these waves encounter them. AM waves can also reflect from a layer of charged particles high in the atmosphere, but FM waves do not.

The waves and wavelengths just mentioned pertain to the carrier waves, which are modulated in amplitude or in frequency according to the amplitude of the sound wave we want to transmit.

The interference by electromagnetic noise with the clean reception of these radio waves is much more severe for AM waves than for FM waves.

Electric current fluctuations need to be amplified to create a large enough effect out of a very small incoming signal. This is done by using the incoming weak signal to regulate the voltage on an obstacle placed in the path of the secondary current flow, since the intensity of this flow is very sensitive to that voltage.

Methods of storing music involve either converting the electric current fluctuations into mechanical motion, which is inscribed onto the surface of a disk, or by using the current fluctuations to partially order the orientation directions of tiny magnets inside magnetic material deposited on the surface of a tape. Tapes tend to have a better response to higher frequencies since the needle of the phonograph, however fine, may be too bulky to be able to follow rapid changes inherent in high-frequency vibrations, whereas the tiny magnets can easily follow such rapid fluctuations. The recent laser-read record was designed to circumvent this difficulty.

The background noise in taped music created by the unevenness and inhomogeneities in the magnetic material of the tape appears mainly at higher frequencies. It can be suppressed by recording the soft parts of the music at higher intensities and then, when replaying, resuppressing the intensity of these soft notes, thus also suppressing the background noise and reinstating the large range of intensities originally present in the music that is taped.

Epilogue

IN THIS BOOK we discussed some of the basic principles of science that allow us to understand music and the activities surrounding it. In doing so, and to make the discussion accessible and enjoyable even to the uninitiated, we had to, on occasion, greatly simplify the situations, to present them in an idealized form. For example, violin strings are not, strictly speaking, one dimensional: they also have a width, and so the wave patterns and frequency rules designed for an idealized one-dimensional case should be somewhat modified in practice. The same is true for wind instruments, which are not strictly one dimensional either. The psychoacoustic phenomena we discussed, such as what is consonant and what is dissonant, are actually more complicated than just a matter of overtones creating beats. When we look at anything in the world, we can first get an approximate idea of how it works through one or a few primary mechanisms, but when interested in the details of its operation to a high degree of accuracy, we have to consider a multitude of different factors that influence the system. Once, however, somebody has acquired a basic understanding of the principles that operate these factors, their application to a system becomes a matter only of diligence, perseverance and motivation.

If we now ask what science is doing to acquire a more accurate understanding of the phenomena discussed in this book, we can answer that a great deal of the work consists of applying these and other principles to realistic situations. For example, research on violin making would have to take into account the fact that the top and bottom of the violin are not flat two-dimensional plates, that their thickness varies over their surface area, that there are imperfections in the material, and on and on.

As we discussed in connection with musical instruments in general, many of the activities surrounding music have proceeded in the past purely on the basis of trial-and-error technology and not the application of scientific principles. We also mentioned that this is changing, and that, especially with the help of computers that do not shy away from lengthy, tedious, and complex numerical calculations based on simple scientific principles, we can handle, in a practical way, increasingly realistic situations. This is one important direction in which current research and development proceeds with regard to music and its surrounding activities.

The second direction is in psychoacoustics. As we stressed in connection with the mechanism of hearing, the perception of music is only partly physical. There is also a psychological factor that is at least as important, which is related to the way in which our brains function. There is much work to be done in this area because, compared with our understanding of physics, our understanding of psychology is very rudimentary.

Finally, the third direction of progress is in the application to music, and to its surrounding activities, of discoveries in science and of the new technologies that develop from them. We have seen many examples of this: virtually everything in Chapters 13 and 14 dealing with the transmission and storage of music is related to such applications of 19th and 20th century science and technology. The radio, the phonograph, the tape recorder, the compact disk, composing on a computer, electronic music, the synthesizer, and acoustic tile are only a few illustrations. This process is ongoing, and as new developments occur in science, they will undoubtedly have an impact also on music. These developments will affect the artistic creation of music also, which in turn has a feedback effect on the type of technological applications evolved for music. The process is a good example of how science and art collaborate in continuing progress.

Mathematics

T HIS BOOK USES only that mathematics one learns in elementary school. It does not involve algebra, trigonometry, or anything higher than that. Nevertheless, for the sake of completeness, this appendix will review those elementary concepts that we do use. These are the four basic arithmetic operations (addition, subtraction, multiplication, and division) as applied to integers and fractions; the concepts of "proportional" and "inversely proportional"; and, very occasionally, powers.

Arithmetic with Integers

The addition, subtraction, multiplication, and division integer numbers are so elementary that they need no review.

Arithmetic with Fractions

A fraction is just a shorthand notation for division. For example, the fraction 15/7 means 15 divided by 7. If you actually carry out this operation manually or on a calculator, you get the decimal fraction 2.1428 . . .

The value of a fraction remains the same if you multiply both the numerator (the 2 in 2/3) and the denominator (the 3 in 2/3) by the same number. Thus 15/7 is the same as 30/14, or 135/63. It is useful to write fractions in the form in which both the numerator and the denominator are as small as possible. In the foregoing example this is 15/7.

To add two fractions, it is necessary to convert them into a form in which the two denominators are the same number. For example, if we want to add 15/7 and 3/4, we first write them in the form of 60/28 and 21/28. Once this is done, we just add the two numerators. In the example this yields 81/28.

The same is true when we want to subtract two fractions. After bringing them into forms in which the two denominators are the same, we subtract the one numerator from the other. Thus $15/7 - 3/4 = 39/28$.

To multiply two fractions, we multiply the two numerators, multiply the two denominators, and then form a fraction out of these two numbers. For example, $15/7 \times 3/4 = 15 \times 3$ over $7 \times 4 = 45/28$.

The best way to divide a fraction by another fraction is to take the second fraction, interchange the numerator and the denominator, and then multiply this new fraction by the first one. Thus, for example, $15/7 \div 3/4 = 15/7 \times 4/3 = 60/21$.

Proportional and Inversely Proportional

A quantity is said to be proportional to another quantity if, when multiplying the first quantity by an arbitrary number, the second quantity is multiplied by the same number. For example, the number of legs in a crowd is (most of the time) proportional to the number of people in that crowd. This means that five people have 10 legs, $4 \times 5 = 20$ people have $4 \times 10 = 40$ legs, and so on.

A quantity is said to be inversely proportional to another quantity if, when *multiplying* the first quantity by an arbitrary number, the second quantity is *divided* by this same number. For example, if a number of partners get together to buy a plot that costs

$100,000, then the cost per partner will be inversely proportional to the number of partners. If only two chip in, the cost per person is $50,000. If three times that many, or six chip in, the cost per person becomes three times less, or $16,667.

It should be noted that "proportional" is sometimes carelessly used to denote simply that if one quantity increases, the other will also. Similarly, this sloppy usage would say that a quantity is inversely proportional to some other quantity if, when the first increases, the second decreases, and vice versa. We will try to stay away from such usage.

Powers

What are powers in the first place? They represent a shorthand notation for a certain kind of multiplication. A number multiplied by itself is said to be that number squared, or that number to the second power. For example, $2 \times 2 = 2^2 = 4$. A number multiplied by itself and then the result multiplied again by the original number is said to be that number cubed, or that number to the third power. Thus $2 \times 2 \times 2 = 2^3 = 8$.

We also use the reverse operation of raising something to a power, namely, the operation of taking roots. The square root of a number is the number that, when raised to the second power, gives the first number. For example, $\sqrt[2]{9} = 3$, because $3^2 = 9$. Similarly, $\sqrt[3]{27} = 3$.

By convention the negative power of a number is 1 divided by that number raised to the same positive number. Thus 3^{-4} means $(1/3)^4$, which is $1/81$.

Powers of 10 often are used because they can simplify notation when we deal with very large or very small numbers. For example, the number 0.000067 can be written as 6.7×10^{-5}.

This powers-of-10 notation is particularly useful for multiplication and division. When multiplying two numbers written in the powers-of-10 notation, we *multiply* the numbers in front of the powers of 10, then *add* the two exponents of the 10, and multiply the two results with each other. For example, $(4 \times 10^5) \times (5 \times 10^2) = (4 \times 5) \times 10^{5+2} = 20 \times 10^7$. Similarly, when we divide two numbers that are written in the powers-of-10 notation, we divide the two numbers in front of the powers of 10 with each other, and attach to the answer 10 raised to the power that is the *difference* of the two original powers. For example, $(4 \times 10^5) \div (5 \times 10^2) = 4/5 \times 10^{5-2} = 0.8 \times 10^3$.

Appendix B
Formulas

As EMPHASIZED IN the text, the approach in this book relies on the understanding of the concepts and simple quantitative relationships that can be used without knowing algebra, and hence without using formulae. Everything in the book is self-contained in that spirit. There are those, however, who either have been corrupted by a different method of science teaching that finds security in memorizing formulas and then "plugging into them," or who have substantial formal skills in physics and mathematics and so can use formulas intelligently. This appendix gives a compendium of the formulas that are implicit in the discussion in the text.

Symbols Used

a acceleration
α absorption coefficient
f frequency
l location (position)
m mass
n positive integer number
r distance
$\varkappa$ reflection coefficient

t time
v speed
A area
C constant
D displacement from the equilibrium position
$\mathcal{D}$ decibel rating
E energy
F force
K kinetic energy
L length of a tube or string
T period
V volume
λ wavelength
μ mass per unit length
σ attenuation coefficient
τ tension

Chapter 1

At times t_1 and t_2, the locations (positions) are l_1 and l_2, respectively, and the speeds are v_1 and v_2 respectively. (In general, l_1 and l_2 are position components and v_1 and v_2, velocity components.) Then for the average speed (velocity component) and average acceleration (component) between t_1 and t_2, we have

$$v_{av} = \frac{l_2 - l_1}{t_2 - t_1}$$

$$a_{av} = \frac{v_2 - v_1}{t_2 - t_1}$$

$$F = ma$$
$$K = \tfrac{1}{2} mv^2$$

Chapter 2

Superposition of two waves, denoted by 1 and 2:

$$D_{tot} = D_1 + D_2$$

Chapter 3

$$v = \lambda f = \lambda/T$$

where v is the speed of propagation of the wave.

Chapter 4

For a sound wave propagating from a point source in all directions uniformly, if the energy per unit area at distance r is E_r, then we have, for two distances r_1 and r_2,

$$\frac{E_{r_2}}{E_{r_1}} = \frac{r_1^2}{r_2^2}$$

For scaling, if the linear dimensions of two similarly shaped objects are r_1 and r_2, and their surface areas and volumes are A_1 and A_2, and V_1 and V_2, then

$$\frac{A_2}{A_1} = \frac{r_2^2}{r_1^2}$$

and

$$\frac{V_2}{V_1} = \frac{r_2^3}{r_1^3}$$

For the Doppler effect, if f is the frequency when the source and observer are at relative rest, and f' is the frequency when there is a relative speed v between the two, then

$$f' = f\left(1 \pm \frac{v}{v_0}\right)$$

where v_0 is the speed of propagation of the wave; $+$ holds for when the source and observer approach each other and $-$ when they move apart.

Chapter 5

$$\mathscr{D} = 10 \; {}^{10}\!\log \frac{E}{E_0}$$

E is the sound wave energy per unit area, and E_0 the minimum sound wave energy per unit area that our ear can perceive (= 10^{-12} watt). ${}^{10}\!\log 2$ is approximately equal to 0.3.

For beats between two waves of frequencies f_1 and f_2, the frequency of the oscillation of the beat wave is ½ $(f_1 + f_2)$, with the amplitude of the beat wave fluctuating with a frequency of $f_2 - f_1$.

Chapters 6–8

No formulas.

Chapter 9

For a tube with both ends open or both ends closed:

$$\lambda = \frac{2L}{n} \qquad n = 1, 2, 3, \ldots$$

For a tube with one end open and one end closed:

$$\lambda = \frac{4L}{2n - 1} \qquad n = 1, 2, 3, \ldots$$

Chapter 10

Frequency of vibration of a string in the fundamental mode with the two ends clamped:

$$f = \frac{1}{2L} \sqrt[2]{\frac{\tau}{\mu}}$$

The frequency of the nth harmonic is $f_n = nf$ where f is the frequency given above.

Chapters 11 and 12

No formulas.

Chapter 13

$$a + \varkappa = 1$$

If E_{Rn} is the sound energy after n reflections, and E_0 the original sound energy, then

$$E_{Rn} = E_0 \varkappa^n$$

For the reverberation time t_R in general, we have

$$t_R = C \frac{V}{A} \frac{1}{-\log \varkappa}$$

The value of C depends on the units used.
For $a \ll 1$, we have the approximate formula:

$$t_R = C \frac{V}{Aa}$$

because

$$-\log \varkappa = -\log (1 - a) = a + \frac{a^2}{2} + \ldots$$

$$\approx a \qquad \text{when } a \ll 1$$

When the wall consists of several segments with different absorprtion coefficients, the approximate formula becomes:

$$t_R = C \; \frac{V}{A_1 a_1 + \ldots + A_n a_n}$$

Attenuation in a medium with attentuation coefficient σ, for the energy E after it traveled distance r in the medium, having been E_0 originally:

$$E = E_0 \, e^{-\sigma r}$$

where e is the base of the natural logarithm, $e = 2{,}71828182\ldots$

Appendix C
Annotated Bibliography

I~N GENERAL YOU~ will find that these books and sources (1) emphasize less the understanding of the basic scientific principles, (2) contain many more encylopedic details, (3) require more sophisticated mathematics, (4) are thicker and more difficult to cover in a one-term or one-semester course by an uninitiated reader. Hence they provide good auxiliary reading, depending on the reader's personality, preparation, interest, and time.

Askill: Physics of Musical Sound. Van Nostrand, 1979. In addition to a discussion of the usual topics (requiring algebra), has an extensive chapter on "hi-fi" and other electronic subjects that provides a qualitative discussion with many clear diagrams. It also has brief sections on electronic music and on ultrasonics, as well as on scales other than the most common ones used in Western music. It includes a description of some experiments pertaining to the subject matter. Exercise problems are offered with their final answers, but without details on how to solve them.

Backus: The Acoustical Foundations of Music. Norton, 1969. A clear exposition by an active researcher in the field. A good bibliography of journal articles for those who intend to go deeper into details. The discussion is comprehensive, but requires algebra for the quantitative aspects. The diagrams are instructive, and the index extensive. No exercise problems.

271

Banade: Horns, Strings and Harmony. Anchor (paperback), *1960.* A qualitative discussion of the subject, covering the ear, and musical instruments, as well as some physical background, but not acoustics, electronics, or the transmission or storing of sound. Has an instructive chapter on "homemade" wind instruments. A short bibliography points to further reading. No exercise problems.

Culver: Musical Acoustics, 4th ed., McGraw-Hill, 1969. An early text with a systematic exposition, some good and detailed diagrams (e.g., page 94, the frequency ranges of various instruments), and a medium-length bibliography. Discussion mainly qualitative with occasional formulas, thus requiring algebra. Exercise questions at the end of chapters but no answers or solutions.

Hall: Musical Acoustics. Wadsworth, 1980. A thick (500-page) book for a one-year course, including sections on usually omitted topics such as the structure of music. The more difficult parts and exercises are marked by asterisks. Many exercises are given, together with their final answers, but the solutions are not worked out. Projects are also outlined. A useful glossary of the many terms used is given. A good reference book.

Rigden: Physics and the Sound of Music, 2nd ed., Wiley, 1985. An unorthodox discussion ranging from pure sound to two pure sounds, to "typical" instruments, to orchestra and acoustics, and to sound reproduction and electronic music. Requires algebra for the quantitative aspects. Good illustrations. Some examples with worked-out solutions. The exercises are provided with final answers but not worked-out solutions. Some references for further reading.

Roederer: Introduction to the Physics and Psychophysics of Music, 2d ed. Springer, 1975. Both more advanced and sophisticated and more limited in coverage than the other books in this bibliography. It does not cover instruments or room acoustics, but focuses on the hearing and perception of music, after an introduction to the underlying physical principles. No exercises. Some excellent diagrams. This book is more a monograph than a text, but serves well for taking the reader farther once the fundamentals have been learned.

Rossing: The Science of Sound. Addison Wesley, 1982. A detailed treatise of more than 600 pages for a full-year course. Covers topics not usually found in texts, such as a chapter on control of noise, and another on measuring instruments. Exercise problems but no solutions. The quantitative discussion requires algebra. A good reference book.

Savage: Problems for Musical Acoustics. Oxford Univ. Press, 1977. A handy paperback with typed text photo-offset, containing many exercise problems on various aspects of the physics of sound, including acoustics, scales, pitch, instruments, and wave propagation. No solutions or answers given.

Scientific American: The Physics of Music. Freeman, 1978. A collection of reprints of articles that appeared in *Scientific American* on various aspects of the physics of music, including the singing voice, piano, woodwinds, brasses, violins, the bowed string, and architectural acoustics. A helpful supplement to knowledge acquired more systematically from a textbook.

White and White: Physics and Music. Holt, Rinehart, & Winston, 1980. A quite detailed (420-page) discussion of the usual aspects of the subject, with many good diagrams, exercise problems (the answers to some are given but not the solutions), examples with their solutions, a medium-length bibliography, and some interesting illustrations of the discussion, such as the Sydney Opera House.

In addition to these books, there is also a collection of reprints of articles from many journals and other sources, entitled Musical Acoustics and published by the American Association of Physics Teachers, Physics Department, State University of New York, Stony Brook, New York 11794, ranging from an original paper by Faraday to contemporary contributions. It details with some special topics in pitch perception, architectural acoustics, and musical instruments.

Exercise Problems

To BENEFIT FROM these problems, work on them first, without looking at the solutions. After spending some time on each of them and going as far as you can, check the solutions in the following section. The proof of understanding in science lies in the ability to use that understanding in problem solving. Thus by working on these problems and then checking how successful you have been in solving them, you can test yourself as to whether you have really understood the ideas presented in this book.

Chapter 1

1.1. A park is 14 miles northwest of a city, while the airport is 14 miles southwest of the city. How far is the park from the airport? (Draw the map and measure.)

1.2. A scientist publishes the following number of papers.

Year	1975	1976	1977	1978	1979	1980	1981	1982
Number of papers	4	6	6	5	9	3	7	5

What is the scientist's average speed of writing papers between 1977 and 1981 (inclusive)? In your answer give both the number and the units in which you are measuring.

1.3. The orbit of a ball with its positions at various times is shown on the graph on page 277. What is the acceleration of the ball (a) in the horizontal direction, and (b) in the vertical direction?

1.4. How much larger is the kinetic energy of a car at 50 mi/h than at 40 mi/h?

1.5. A rocket is launched from Cape Canaveral. Its position is described, at various times after the launch, by the following table.

Time (seconds)	North–South Position (miles)	East–West Position (miles)	Height (miles)
0	0	0	0
10	1.0	2.5	1.2
20	3.1	7.8	3.5
30	6.2	15.4	8.9
40	10.1	28.3	15.4
50	15.8	47.6	29.1

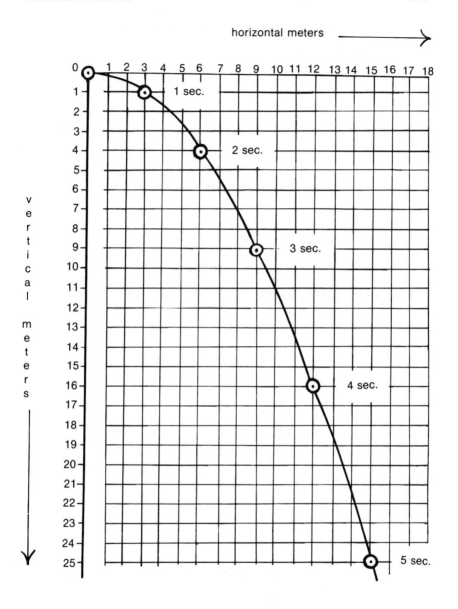

horizontal meters ⟶

Determine the following quantities.

a. The average north–south speed in the first 20 seconds.

b. The average east–west acceleration in the time interval be-tween 20 and 40 seconds.

c. The vertical speed averaged over the first 50 seconds.

1.6. An apple separates from the tree branch and falls to the ground. Describe this event in terms of energy-conversion processes.

1.7. You crash two identical cars into a heavy, thick wall, one at 20 mi/h, the other at 40 mi/h. Is the damage to the second car likely to be twice as much as to the first car?

1.8. You leave Eugene, Ore., at 12 noon. Your initial speed is 0.80 mile per minute north, and 0.25 miles per minute west. You then proceed at a steady acceleration of 0.05 mile per minute^2 north and 0.02 mile per minute^2 west. What will be the north–south and the east–west locations after 10 minutes?

Chapter 2

2.1. The table below gives the distance of the piston in a car from the end wall of the cylinder of the car engine. Determine the period, frequency, and amplitude of the piston's oscillation in the cylinder.

Time (sec)
   0.025 0.030 0.035 0.040 0.045 0.050 0.055 0.060 0.065 0.070
Distance (cm)
    11   17   20   21   20   17   11   5   2   1

Time (sec)
   0.075 0.080 0.085 0.090 0.095 0.100
Distance (cm)
    2   5   11   17   20   21

2.2. The inflation rate in a country in consecutive years is as follows:

Year	1950	1951	1952	1953	1954	1955	1956	1957
Inflation rate	3.8%	4.8%	5.2%	6.1%	5.4%	4.5%	4.0%	4.3%

Year	1958	1959	1960	1961	1962	1963	1964
Inflation rate	5.0%	5.8%	5.3%	4.7%	4.1%	4.9%	5.3%

What are the approximate values of the period and the frequency of the inflation oscillations?

2.3. You go to the amusement park, and just as you get there at 2:30 p.m., a train of the roller coaster leaves the station. When you return later at 4:10 p.m., the little train has returned to its station five times and is just leaving on its sixth ride. What are the frequency and period of the motion of the roller coaster train?

2.4. What is the phase difference between the two waves shown in the following?

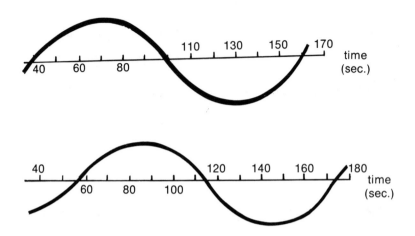

2.5. On p 281 you have the graph of two oscillations. Underneath, draw the superposition of these two oscillations.

2.6. Construct the graph of the oscillation for which the Fourier recipe is as follows:

amplitude

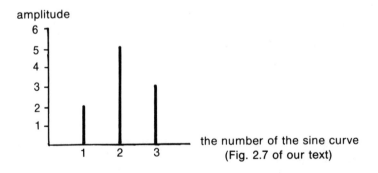

the number of the sine curve
(Fig. 2.7 of our text)

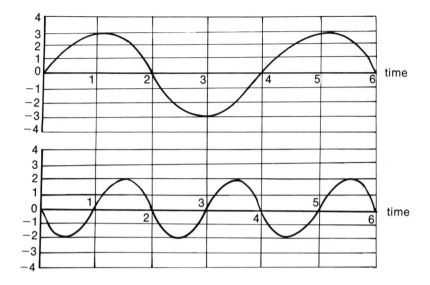

Chapter 3

3.1. Walking is a periodic motion. If you take 50 steps every 100 feet, what is the wavelength of your walk?

3.2. You float at a given spot in the ocean and find that it takes three seconds between being on the crest of a wave and being in the next trough. At the same time, somebody from a helicopter surveys the waves from above and finds that the two neighboring crests are 18 feet apart. What is the speed of propagation of the wave?

3.3. Determine the wavelength, period, frequency, and speed of propagation of the wave shown in the following.

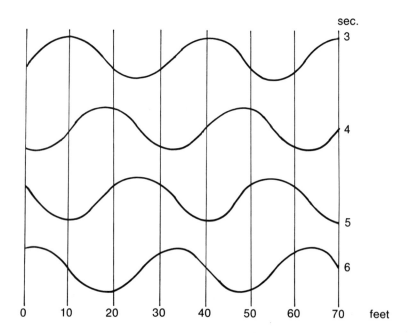

3.4. I am floating in the ocean, with a stopwatch in my hand. I therefore can determine that four seconds pass between my being on the crest of the wave two consecutive times. When I am on the second crest, I look toward the shore and see that the first crest I was on four seconds ago is now 15 feet away from me.

a. What is the frequency of the ocean wave?

b. What is the wavelength of the ocean wave?

c. What is the speed with which the ocean wave travels?

3.5. A wave with an amplitude of 10 units, and another wave of the same frequency with an amplitude of four units and 180 degrees out of phase with the first wave, are superimposed. What is the amplitude of the resultant wave, and what is the phase difference between the first wave and the resultant wave?

3.6. An ocean wave travels at a speed of 5 ft/s and the neighboring crests of the wave are 35 feet apart. Floating in the ocean at a given spot, you are on a crest at a given time. How much time will elapse before you are on a crest the next time?

3.7. At the top of p 284, are representations of two waves *A* and *B*. In the graph showing *A*, draw the wave *C*, which, if superimposed on wave *A*, would yield a resultant wave like *B*. Indicate the amplitude and wavelength of your wave *C*.

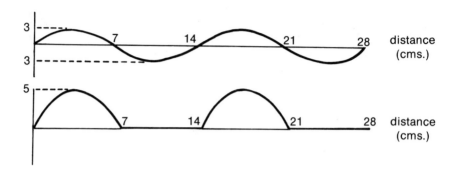

3.8. Sound waves in air are incident on a surface of water, as shown in the figure. The straight lines (dashed) represent the position of the crests of the waves at a given time. Some of the waves will refract and enter the water. Draw in the corresponding lines for the refracted wave.

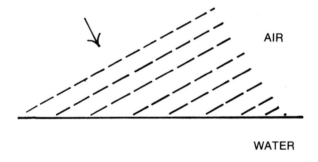

3.9. I put my ear to the railroad rail and ask a friend to transmit the musical note A (frequency 440 per second) to the rail at a place 2500 meters from my ear. I hear this sound 0.50 seconds later. What is the wavelength of that sound wave in the rail?

3.10. I emit one wave at point *A* and another at point *B*. The first has a wavelength of two meters, and the second a wavelength of three meters. They have the same amplitudes. I want to choose the distance between *A* and *B* so that I get complete destructive interference at *C* at all times. The distance from *B* to *C* is four meters. What must the distance between *A* and *B* be in order to accomplish the stated aim?

Figure:

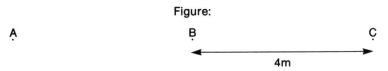

Chapter 4

4.1. A car horn is sounded as the car approaches me, passes by, and then recedes. I hear the lowering of the pitch of the horn. How fast would the car have to go for this lowering of the pitch to be by 50 per second in frequency. Can an answer be given on the basis of the preceding information? If not, what else would be needed?

4.2. You recall the example of the Doppler effect involving the young man, the woman, and the letter-carrying pigeon. At what time on day 4 would the man have received the letter had he been walking *away* from the woman instead of toward her?

4.3. The first act of the opera performance is about to close. The tenor is running from the rear of the stage toward the soprano at the front of the stage, with outstretched arms, at a speed of two meters per second (a remarkably agile tenor!). He is singing "amore!" on a high C note, which has a frequency of 523 per second. The tenor, by his own judgment, sings the note on pitch. Will the audience hear it so, and if not, in what way will the audience hear it differently, and how differently?

4.4. A 10-inch pizza (i.e., one with a diameter of 10 inches) costs $2.00, while a 12-inch pizza costs $2.70. What is a better deal?

4.5. An auditorium of a given size is mandated to have eight emergency fire exits. There is a second auditorium, twice as long and twice as wide, but equally tall as the first one. How many emergency exits should this second auditorium have?

4.6. A steady sound source is located at point *A*. I have a measuring instrument that registers the amount of sound energy per unit area at a given location *B*, which is 3.4 miles from *A*. The instrument indicates 35 in its own units of energy per unit area. I then take the instrument to another point *C*, which is 5.1 miles from *A*. What will the instrument register there? Does the answer depend on circumstances not given in this description?

Chapter 5

5.1. The noise on the ground of a plane flying overhead at 30,000 feet is 42 dB. Approximately how many decibels would the noise of the same plane be if it flew at 15,000 feet?

5.2. A trombone plays a note that is recorded to be at a loudness level of 60 dB. About how many trombones would have to play simultaneously (and each as loud as the first one) to record a loudness level of 66 dB?

5.3. What can you say about the two waves the superposition of which produces the following wave?

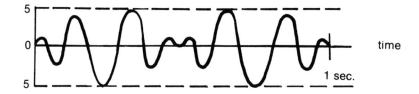

5.4. A town has an emergency siren costing $100, which, at the city limit, is heard at 60 dB. This is judged not sufficiently loud, and so the city decides to spend $700 more to buy seven more identical sirens and install them at the same location where the first siren is located. At what decibels would you hear the resulting total arrangement of sirens when you are at the city limit?

5.5. A certain sound outdoors is heard at a distance of 200 feet from the source at 60 dB. How far do I have to move from the sound source so that I hear the sound at 45 dB?

5.6. A certain number of tubas play together and each is equally loud. Together they produce, at a given location for the observer, 82 dB of intensity. Now two of the tubas stop playing while the others continue. Consequently the sound level drops to 79 dB. How many tubas played together originally?

5.7. Two "pure" sounds of equal intensity, one at a frequency of 200 per second and the other at a frequency of 204 per second, are sounded together. As a result I will hear a sound. Describe what this sound would be like.

5.8. An explosion of an antiaircraft artillery shell in midair produces 69 dB at an observer one kilometer from the explosion. How many decibels will another observer register who is 10 kilometers from the explosion?

5.9. A thunder clap is heard by somebody at 60 dB. Somebody 4½ miles farther from the lightning hears the same thunder clap at 40 dB. How far was the lightning from the first observer?

Chapter 6

6.1. Take a piece of music you know well, and analyze it in terms of the five different time scales we discussed.

6.2. I sound a note on a piano, tape the sound, and then play the tape backward. Draw a graph of the sound wave, as a function of time, of this backward piano sound.

6.3. Your own life can also be described in terms of a nested set of time scales. Do so.

Chapter 7

7.1. Is the musical interval between the second and the sixth notes of the tempered scale larger or smaller than a diatonic fifth?

7.2. The frequency of a certain C is 262 per second. Calculate the frequency of the F a tempered fourth above it.

7.3. The standard A is 440 per second. How many beats a second would I hear if I sounded that A, together with the A which is a diatonic fifth above the D which is a tempered fifth below the 440 per second A?

7.4. Starting at a note, I go up a diatonic fifth and, using that as a base, I go up a diatonic fourth. I compare this final note with a note I get by going up an octave from the original note. What is the result of this comparison?

7.5. A well-tuned piano and an old, seldom-tuned piano are side by side and the middle As on the two instruments are sounded together. (This middle A is supposed to have a frequency of 440 per second.) An unpleasant sound with four beats per second is heard. In terms of *musical intervals*, by how much is the old piano off pitch?

7.6. I want to construct a new type of a scale, which is tempered and has only four (and not six) notes between the two notes of an octave. What would be the frequency ratios of the six notes of this new scale with respect to the note at the lower end of the octave? (*Hint:* $1.15^5 = 2.00$)

Chapter 9

9.1. A musical instrument in which the secondary vibrator is a tube with both ends open produces a D a diatonic fifth above the standard 440 A. How long is the tube?

9.2. A trombone produces an A two octaves below the standard 440 A. This note is produced in the fundamental mode, that is, it is the lowest frequency that can be produced on a tube of that length. I now want to raise the pitch by one tempered semitone (that is, from A to the tempered A-sharp). How much do I have to move the hand that holds the sliding part of the trombone to do so?

9.3. A valveless bugle has a tube 0.5 meter long. What are the frequencies of the notes this bugle can produce?

9.4. A valved brass instrument with none of the valves activated (that is, the air passage being the shortest possible distance) has an air passage length of two meters. When one of the valves is activated, and thus one of the bypasses is inserted into the air passage, the length of the air passage becomes 2.12 meters. By how much will the musical pitch of the instrument change when I activate that valve? Will it rise or fall? Will the musical interval of the change be different if I play in the fundamental mode than if I overblow and play in the first harmonic mode?

Chapter 10

10.1. A sound played on the empty (untouched) string of a violin has the spectrum shown at the top of p 292. Name at least one position along the string where this sound could *not* have been bowed.

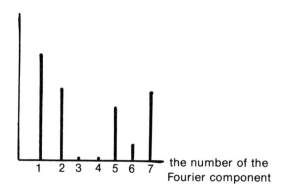

the number of the
Fourier component

10.2. A vibrating string is in a mode in which there are three extra nodes between the two ends. It vibrates with a frequency of 1200 per second. What is the approximate frequency of its fundamental vibration mode (i.e., the one with no extra nodes between the ends)?

10.3. New York City's Lincoln Tunnel under the Hudson River is 1800 meters long. Viewed as a wind instrument, what would be the frequency of its vibration patterns with the longest wavelength? Consider instead a rope stretched across the Hudson River, above it, in the air, also 1800 meters long. How would its vibration frequency compare with the frequency of the Lincoln Tunnel?

10.4. The lowest note of the violin is G (with a frequency of 196 per second), while the double bass has a lowest note of an E (with a frequency of 41.2 per second). If the string of the violin is 40 centimeters long, how long would the double bass string have to be with the same diameter and under the same tension?

10.5. The viola is 1.17 times as long as the violin, and yet the lowest frequency of the latter is 1.5 times larger than the lowest frequency of the former. How is this possible?

Chapter 11

11.1. A drum has a fundamental frequency (i.e., the lowest frequency it can produce) of 50 per second. I want to excite the 12th harmonic above this fundamental. Should I use a mallet with a large or small head?

11.2. I hit the surface of a kettledrum (circular) right at the center. Which of the vibrational patterns shown in Figure 11.1 will be excited? What will be the (approximate) musical intervals between the first three harmonics? (Use the parenthetical frequency values.)

Chapter 12

12.1. The speed of sound in helium is 980 meters per second. By what musical interval will the helium-filled singer's voice rise in pitch compared with the normal, air-filled state?

12.2. By experimentation on yourself, determine the difference between "p" and "f" from the point of view of sound production.

12.3. By experimentation on yourself, determine the difference between "a" as in "law" and "a" as in "made," from the point of view of sound production.

Chapter 13

13.1. We have a long, narrow, and low auditorium, as shown in the illustrations. The four long rectangular sides of the hall are covered with material with an absorption coefficient of 1.0. The

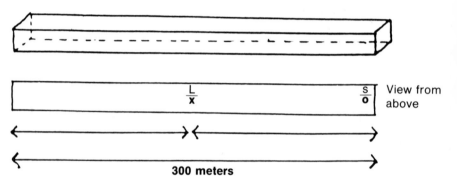

two small square-shaped ends of the hall are covered with a material with an absorption coefficient of 0.5. The source S, at one end of the hall, produces a loud and sharp (very brief) sound. Halfway down the hall, there is a listener L. On the graph indicate the sound intensity at various times that the listener will hear. How much is the initial time delay in this example? How much is the reverberation time?

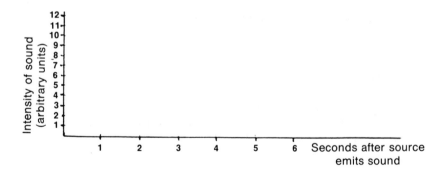

13.2. A wall looks like that in this illustration. Sound waves hit the wall from the direction as shown.

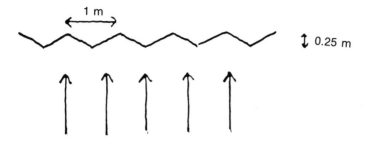

a. If the frequency of the impinging sound wave is 100 per second, in what direction will the reflected waves come out?

b. What would be the answer if the frequency of the waves were 3000 per second?

13.3. A concert hall is cube shaped, 50 meters on the side, and all of its inside surfaces are covered with a material that has a reflection coefficient of 0.60. Its reverberation time is 3.4 seconds. If the concert hall were a cube 100 meters on the side, but covered with a material with a reflection coefficient of 0.36, what would be its reverberation time?

Chapter 14

14.1. The response curve of an amplifier is as shown.

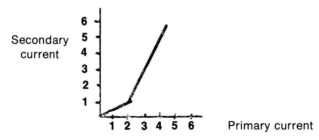

I feed into this amplifier a primary current pulse with the following shape.

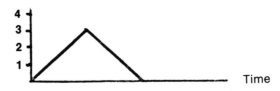

What will the secondary current pulse look like?

14.2. Suppose a tape recording of some music is made at a tape speed of 19 centimers per second, and then replayed at **a** 6⅓ centimeters per second tape speed. By what musical interval would the pitch of the music be changed, and in what direction?

14.3. Every so often there is a big explosion on the sun, which results in the emission of an intense stream of charged particles, and when this particle beam hits the earth's atmosphere, it disrupts the charged particle layer that is usually in the earth's atmosphere at a certain altitude. What effect will such an event have on radio reception in the AM and FM bands?

Solutions to Exercise Problems

Chapter 1

1.1.

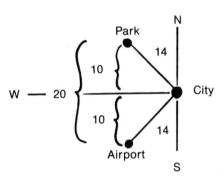

1.2. $\dfrac{6 + 5 + 9 + 3 + 7}{5} = \dfrac{30}{5} =$ six papers per year

1.3.

Horizontal direction:

Time	0		1		2		3		4		5	sec
Distance	0		3		6		9		12		15	m
Speed		3		3		3		3		3		m/sec
Acceleration			0		0		0		0			m/sec^2

Vertical direction:

Time	0		1		2		3		4		5	sec
Distance	0		1		4		9		16		25	m
Speed		1		3		5		7		9		m/sec
Acceleration			2		2		2		2			m/sec^2

1.4. The kinetic energy goes as the *square* of the speed. If, therefore, the ratio of speeds is 50/40 = 1.25, then the ratio of the kinetic energies is $(1.25)^2$ = 1.56. So the kinetic energy at 50 mi/h will be 56 percent more than it was at 40 mi/h.

1.5.

a. (3.1 − 0)/20 = 0.155 mile/second

b. Speed (east–west direction) between 20 and 30 seconds: (15.4 − 7.8)/10 = 0.76 mile/second

Speed (east–west direction) between 30 and 40 seconds: (28.3 − 15.4)/10 = 1.29 miles/second

So the acceleration is (1.29 − 0.76)/20 = 0.0265 mile/second^2

This is pretty rough, because we can calculate only the *average* speeds between 20 and 30, and between 30 and 40.

c. 29.1 miles over 50 seconds gives an average vertical speed of 0.582 miles/sec. (29.1/50).

1.6. As the apple separates, it has zero speed, but has some potential energy since its position is above the ground. As it drops, its potential energy decreases because its height above the ground decreases, but it picks up speed so that its kinetic energy increases. As it hits the ground, its potential energy becomes zero, and its kinetic energy is converted into heat and into potential energy of its molecules as the apple is deformed or breaks into pieces.

1.7. Since the damage is caused by the kinetic energy of the car being converted into the potential energy of molecules acquired in the deformation or breakup of the body of the car, the damage will be *four* times larger when the speed is increase by a factor of *two*.

1.8. Let us prepare the table for the distance, speeds, and accelerations for every minute. First north:

Time:	0	1	2	3	4	5	6	7	8	9	10
Location:	0	0.80	1.65	2.55	3.50	4.50	5.55	6.65	7.80	9.00	10.25
Speed:	0.80	0.85	0.90	0.95	1.00	1.05	1.10	1.15	1.20	1.25	1.30
Acceleration	0.05	0.05	0.05	0.05	0.05	0.05	0.05	0.05	0.05	0.05	0.05

Now west:

Time:	0	1	2	3	4	5	6	7	8	9	10
Location:	0	0.25	0.52	0.81	1.12	1.45	1.80	2.17	2.56	2.97	3.40
Speed:	0.25	0.27	0.29	0.31	0.33	0.35	0.37	0.39	0.41	0.43	0.45
Acceleration:	0.02	0.02	0.02	0.02	0.02	0.02	0.02	0.02	0.02	0.02	0.02

The times are in minutes, the speeds in miles per minute, the locations in miles, and the accelerations in miles per minute squared.

The north–south location, therefore, after 10 minutes, is at 10.25 miles north, while the east–west location is at 3.40 miles west.

The calculations are only approximate, since the speed changes are calculated only every minute, while in reality the acceleration (change of speed) occurs gradually.

Chapter 2

2.1. Either by looking at the numbers or by plotting them on graph paper, one can see that the period is about 0.060 second, hence the frequency of $1/0.060 = 16\frac{2}{3}$ per second. We also see that the distance changes between one and 21 centimeters, so the amplitude is $\frac{1}{2} (21 - 1) = 10$ centimeters.

2.2. By plotting the figures, or even by just looking at them, one sees that the period is about six years, and hence the frequency is 0.167 per year.

2.3. If the train made five rounds in one hour and 40 minutes, or 100 minutes, its period is 20 minutes, and thus its frequency is three per hour.

2.4. The period of the wave, as one can read off the graph, is 160 − 40 = 120 seconds. The lower wave is 15 seconds "behind" the upper one. The phase difference is expressed by equating the one period to 360 degrees, which, in this case, is 120 seconds. Thus a 15-second delay of the second wave corresponds to (360/120)15 = 45 degrees phase difference.

2.5.

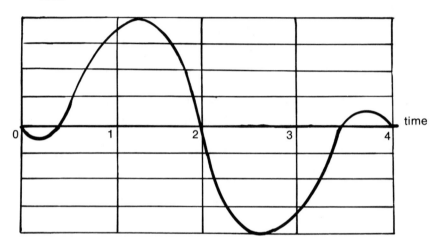

2.6.
(See top of page 301).

Chapter 3

3.1. Each step is two feet. One period of the walking motion is one step each with your two feet. Hence the wavelength is four feet.

3.2. The period is three seconds, and the wavelength is 18 feet. The speed of propagation of the wave is therefore 18 feet/three seconds = 6 feet/second.

3.3. The wavelength can be directly read off any of the four curves, for example, by taking the distance between two consecutive crests. It is 30 feet. The period can be determined by looking at all four pictures together. For example, you see that the first and

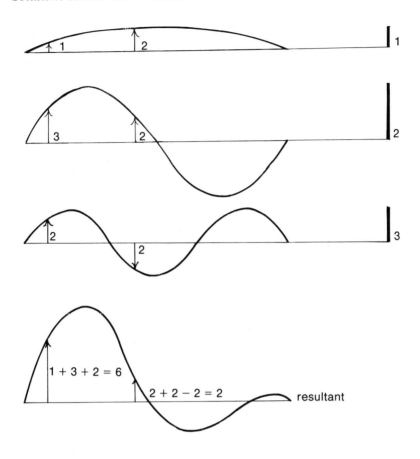

the third waves (at three seconds and five seconds, respectively) are completely (i.e., by 180 degrees) out of phase. In another two seconds, therefore, the wave would again be in phase with the one at three seconds. So the period is four seconds. From this the speed of propagation is 30 feet/four seconds = 7.5 feet/seconds.

3.4. This is very similar to 3.2. The period is four seconds, the wavelength is 15 feet, the frequency is one-fourth per second, and the speed is 3.75 feet/second.

3.5. By drawing the two waves and then constructing the result, you see that the resultant amplitude is six units and the resulting wave is exactly in phase (i.e., with 0 degrees phase difference) with the first wave.

3.6. Since the speed of propagation is five feet/second, and the wavelength is 35 feet, therefore the period must be seven seconds, or the frequency 0.14 per second. So seven seconds will elapse before you are on the crest again.

3.7.

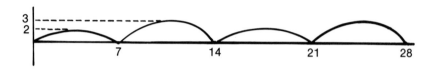

3.8.

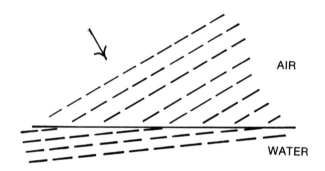

3.9. The sound in the rail covers 2500 meters in a half a second, and hence its speed is 5000 meters/second. For a wave with a frequency of 440 per second, the wavelength must be (5000 meters/second)/440 per second = 11.36 meters.

3.10. It is impossible to choose C to achieve the stated aim. Since the two waves, traveling in the same medium (and hence presumably with the same speed), have different wavelengths, they will also have different frequencies, and hence nowhere will the two waves cancel each other *at all times*.

Chapter 4

4.1. The *ratio* of the change of frequency to the original frequency (at relative rest) is the same as the ratio of the relative speed of motion to the speed of sound. Hence to calculate the desired speed of the car, we would have to be given also the original frequency of the car horn, that is, the frequency heard by somebody not moving with respect to the car. For example, if this original frequency were 400 per second, then we would have

$$\frac{50 \text{ per second}}{400 \text{ per second}} = \frac{\text{The desired speed of the car}}{600 \text{ mi/h (which is the speed of sound in air)}}$$

from which we could get the desired speed to be 75 mi/h.

4.2. Instead of meeting the pigeon eight hours earlier than noon, he would have met it eight hours later, that is, at 8 p.m.

4.3. This is similar to problem 4.1. Thus we have

$$\frac{2 \text{ meters per second}}{300 \text{ meters per second}} = \frac{\text{Change of frequency in units of per second}}{523 \text{ per second}}$$

The change of frequency would therefore be about 3.5 per second, or about 0.6 percent, which is too small to be perceived by the human ear.

4.4. The value in the pizza is in the amount of food, which depends on the area of the pizza (assuming that small and large pizzas have equal heights). Thus the ratio of values in a 10-inch and 12-inch pizzas is

$$\frac{12^2}{10^2} = 1.44$$

Thus if the smaller pizza costs $2.00, the larger would give you the same value at $2.00 × 1.44 = $2.88. Since it actually costs only $2.70, it is a better buy than the small one.

4.5. The number of people in the two auditoriums is proportional to the area of each auditorium. Thus the larger one will have $2 \times 2 = 4$ times as many people in it. If, therefore, you want to be able to evacuate the large auditorium as rapidly as the small one, you need in the former $8 \times 4 = 32$ emergency exists.

4.6. We know that the sound energy decreases as the square of the distance. Thus we have the relationship

$$\frac{3.4^2}{5.1^2} = \frac{\text{The reading on the meter at the larger distance}}{35}$$

or we get about 15.6 for the reading on the meter at 5.1 miles from *A*.

Chapter 5

5.1. Twice closer means that the noise will be $2^2 = 4$ times more. Each multiplication by 2 increases the decibel rating by 3. So the 42 dB will increase to $42 + 3 + 3 = 48$ dB.

5.2. Sixty-six decibels is 6 dB more than 60, and each 3 dB means a multiplication by 2. So instead of the original one trombone, we would have four.

5.3. The pattern looks like a beat. The frequency of the "wobble" of the beat is two per second, which is then the *difference* of the two superimposed frequencies. The frequency of the wave itself is eight per second (you can see eight peaks in the one-second interval). This is the *average* of the two original frequencies. So the pattern shown is the superposition of two waves of equal amplitudes, one with a frequency of nine per second and the other with seven per second.

5.4. The resulting eight sirens are 2^3 times what they were before the new purchase. Each multiplication by 2 means an addition of 3 to the decibel rating. So the final decibel level will be $60 + 3 + 3 + 3 = 69$ dB.

5.5. The difference between 60 and 54 is 6 dB. That means a decrease of sound intensity by a factor of four. Since the intensity decreases as the square of the distance, such a decrease corresponds to making the distance twice as much, or 400 feet.

5.6. The decrease from 82 dB to 79 dB is 3 dB, which means a decrease in the sound intensity by a factor of two. If such a decrease corresponds to two tubas dropping out, there had to be four tubas playing to begin with.

5.7. There will be a beat, with a "wobble" four times a second, and a wave frequency of 202 per second.

5.8. The sound intensity drops as the distance squared. Thus increasing the distance by a factor of 10 means a decrease in sound intensity by a factor of 100. Each division by 10 means a drop in the decibel rating by 10. Thus the final decibel rating will be $69 - 10 - 10 = 49$.

5.9. The decrease is by 20 dB, which corresponds to a decrease in sound intensity by a factor of 100, which in turn means an increase in distance by a factor of 10. Hence the lightning took place at a distance of 4.5 miles/10 = 0.45 mile from the first observer.

Chapter 6

6.1. Depends on what your favorite piece of music is. Have fun with it.

6.2. Since the piano sound is a composite of a fundamental frequency and overtones (Fourier components of various kinds), it will have some irregular but periodic shape. In addition, the sound is strong when you hit the key and then quickly (within a second or so) fades. Hence the actual waveform of the note would be something like the picture on the top of page 306.
If you play it backward, you get this same picture going from right to left.

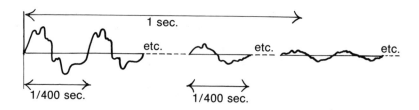

6.3. Your heart beats about every second. You have a daily routine of getting up, eating, sleeping, etc. You have a weekly routine of a schedule of classes or work. If a student, your life is also periodic over a three-month period, taking classes and taking exams. You have a yearly cycle of the seasons, and if a student, of your academic year. Finally, there is the human life cycle of being born, growing up, and dying.

Chapter 7

7.1. The ratio of frequencies for the diatonic fifth is 1.5. The ratio between the second and sixth notes in a tempered scale is

$$\frac{1.6818}{1.1224} = 1.4983$$

So we see that the interval in the diatonic fifth is larger.

7.2. Any tempered fourth represents a ratio of 1.3348. Thus the frequency of the desired note is $262 \times 1.3348 = 349.7$ per second.

7.3. If we go a tempered fifth below the 440, we get a frequency of

$$\frac{440}{1.4983} = 293.67$$

If we now go a diatonic fifth up from this note, we get $293.67 \times 1.5 = 440.50$ per second

If we then sound the two notes together—namely, 440 and 440.50—we get one beat every two seconds.

7.4. Going up a diatonic fifth means multiplication by 3/2. Then, going up a diatonic fourth from there means a further multiplication by 4/3. Since

$$\frac{3}{2} \times \frac{4}{3} = 2$$

we get to the same place we would have by raising the first pitch by an octave.

7.5. Four beats per second corresponds to a difference in frequency of four, which means that the two pianos have A's that are 440 and 444 respectively. The ratio of 444 and 440 is 1.0091, and since 1.06 is approximately a semitone, the difference between the two A pitches is about one-sixth semitone.

7.6. The frequencies of the six notes (including the ones at the bottom and the top of the octave) will be

1.00   1.15   1.3225   1.5209   1.7490   2.00

times the frequency of the note at the bottom of the octave.

Chapter 9

9.1. The diatonic D above the 440 A has a frequency of (3/2) times 440, or 660 per second. The wavelength of that sound is 330 meters/second divided by 660 per second, or 0.50 meter. The length of a tube open at both ends can be half the wavelength, the wavelength itself, three halves times the wavelength, twice the wavelength, etc. Hence the length of the instrument that can produce the required sound can be 0.25 meter, 0.50 meter, 0.75 meter, 1.0 meter, etc.

9.2. The note two octaves below the 440 A has a frequency of one-quarter times 440, or 110 per second. Its wavelength is, therefore, 300 meters/second divided by 110 per second, or 2.72 meters. A tempered semitone higher note has a wavelength about 6 percent shorter, or 0.16 meter shorter. Thus I have to move my hand by half that distance, or 0.08 meter (about three inches).

9.3. The bugle is an instrument with one end open and one end closed. Thus the relationship between the wavelength and the length of the instrument is that the wavelength is four times the tube length, or one-third times that, or one-fifth times that, or one-seventh times that, etc. Thus in this case the wavelengths will be two meters, 0.67 meter, 0.4 meter, 0.28 meter, etc., and the corresponding frequencies of 150 per second, 450 per second, 750 per second, 1050, per second, et.c

9.4. The change in tube length from two meters to 2.12 meters is 6 percent or just a tempered semitone. It makes no difference whether I play the fundamental mode or some higher harmonic of it; the percentage change is the same. The tone with the valve inserted will be lower.

Chapter 10

10.1. The shapes of the standing waves corresponding to the Fourier components are given below. Since in the spectrum the third and fourth components are missing, but the others are present, we could not have bowed in a position that is a node for any that are present. Thus, for example, we could not have bowed in the middle of the string, because that is a node position for the second component.

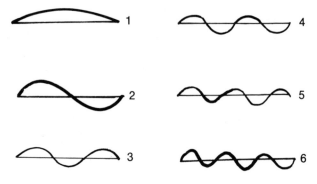

10.2. The two vibrational modes are pictured here. We see that the wavelengths have a ratio of 4:1, and hence the frequencies 1:4. The frequency of the fundamental must be 1200/4 = 300 per second.

10.3. For a tube with both ends open, the longest wavelength is twice the length of the tube, or, in our case, 3600 meters. Thus the frequency would be 300 meters/second: 3600 meters = 1/12 per second, which, of course, is not audible sound.

If we stretch a rope across the river, with both ends fixed, the wavelength of the wave with the longest wavelength would also be twice 1800 meters, or 3600 meters. The corresponding frequency on the rope, however, would depend on the tension on the rope and on the thickness and material of the rope, and hence we cannot calculate the frequency without additional information.

10.4. If the tension on the strings and their thickness are the same, then the frequency is just inversely proportional to the length. Since the ratio of 196 to 41.2 is 4.75, the length of the double-bass string would have to be 40 × 4.75 = 190 centimeters.

10.5. It is possible, because for string instruments length is not the only determinant of the frequency. So in this case, we can produce such a lower frequency on the viola by putting the string under less tension and/or making the viola string out of thicker and/or heavier material.

Chapter 11

11.1. I should use the mallet with the small head, since high haromics have vibrational patterns with many wiggles, and are excited easier by a mallet that hits the drum surface in as small an area as possible.

11.2. If I hit the drum surface at the center, the center cannot be along a node line. Hence all patterns shown in Figure 11.1 are excluded except for the three in the leftmost column of that figure.

Chapter 12

12.1. Since the speed of sound in helium is about three times larger than the speed of sound in air, the frequencies will also rise by a factor of about three. A factor of two raises the pitch by one octave, and an additional factor of 1.5 raises the pitch another fifth. So, approximately, the pitch will rise about an octave and a half, that is, an octave and a fifth.

12.2. You will find that the obstruction placed in the path of the outgoing air in the case of "p" is made by your lips, while the obstruction for "f" is made jointly by your lips and teeth.

12.3. You will find that in forming the "a" in "law" more of the back part of your mouth is involved, and you form your lips into a small circular hole. In contrast, the "a" in "made" is formed more in the front, against your palate, and you open your mouth more broadly.

Chapter 13

13.1. The graph would look like this.

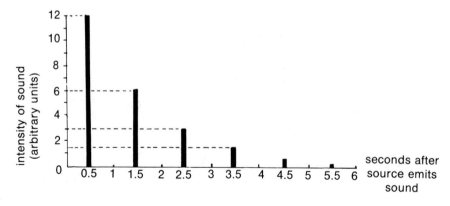

The initial time delay is 1.0 seconds.

In principle, the reverberation time would be about 20 seconds, since it would take 20 reflections to decrease the sound intensity by 60 dB or 10^6. This is so since each reflection decreases it by 2, and 2^{20} is about 10^6.

13.2. For a frequency of 100 per second, the wavelength is meters/second divided by 100 per second, or three meters, which is long compared with the size of the corrugations on the wall. Thus such a wavelength would see the wall flat, and hence the waves would reflect in the same direction from which they came. For a frequency of 3000 per second, the wavelength is 0.10 meters, small compared with the dimensions of the corrugations in the wall, and so the waves would reflect as indicated here.

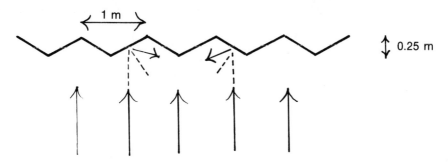

13.3. Raising the linear dimension of the hall would, other things being equal, raise the reverberation time by the same factor, or in this case double it. But making the reflection coefficient the square of the previous reflection coefficient, on the other hand, would, other things being equal, decrease the reverberation time by a factor of two. So when both changes are made simultaneously, the reverberation time remains the same.

Chapter 14

14.1. See figure on page 312.

14.2. The speed is decreased by a factor of three, so all frequencies will also drop by the same factor. A factor of two is an octave, and then a factor of 3/2 is another major fifth, so the pitch will drop by an octave and a fifth, or about 1½ octaves.

14.3. Since FM does not bounce back from the charged particle layer, but AM does, one would expect much more disruption in AM communication than in FM.

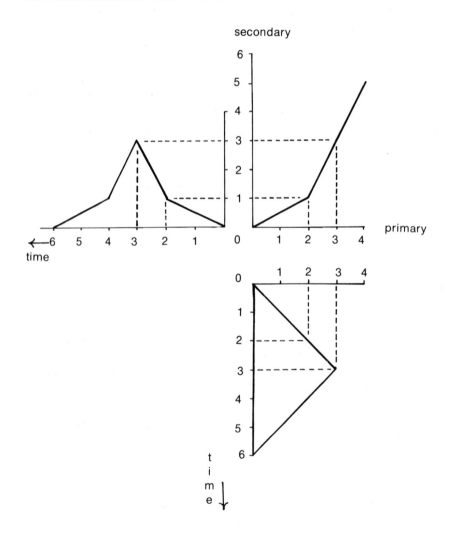

Index

Michael J. Moravcsik has performed the unique feat of explaining the science of musical sound to readers with no knowledge of mathematics beyond the sixth grade of elementary school. The result is a sparkling, lucid work that will appeal to the music lover at large as well as to the college or high school student of music or physics. He communicates important scientific ideas to people who have no formal scientific background.

The major physical ideas pertaining to waves and other phenomena relevant to the study of sound are developed from the beginning. Concepts like acceleration, force, energy, resonance, vibration, frequency, wave length, destructive interference, Fourier decomposition emerge simply and clearly.

The physical background is then used to describe the traveling of sound waves, the way people hear sound, the features of musical sound, musical pitch, and musical scales.

The discussion then turns to musical instruments. The author shows that all instruments resemble each other in their main components. These are then discussed in detail for wind, string, percussion instruments, and the human voice.

Finally, a clear exposition is given of the acoustical aspects of the environment of music, and the transmission and storage of sound through radio, phonograph, tape recorder, and other equipment.

A brief appendix summarizes the mathematics used in the book. Although algebra is not used in the text, a second appendix summarizes those formulas related to the content of the book. There is an annotated bibliography, as well as a large number of exercise problems, with their detailed solutions. The book is illustrated with charts, graphs, and photos. *Fully indexed.*